NICK BREEZE has been interviewing climate scientists and related experts since setting out in 2009 with a film crew to interview Professor James Lovelock for a proposed documentary with the working title A *Hitchhiker's Guide to Gaia*. In 2017, he co-founded the Cambridge Climate Lecture Series (CCLS), which brings world-renowned experts to the University of Cambridge for livestreamed lectures on a range of climate issues. Nick is a wine journalist, and the UK Ambassador for the Wines of Alentejo Sustainability Programme.

T0020825

COPOUT

How governments have failed the people on climate

NICK BREEZE

An insider's view of Climate Change Conferences,
from Paris to Dubai

First published in the UK in 2024 by Ad Lib Publishers Ltd
Marine House, Tide Mill Way,
Woodbridge, Suffolk IP12 1AP
www.adlibpublishers.com

Text © 2024 Nick Breeze

Paperback ISBN 9781802472059
eBook ISBN 9781802472196

A CIP catalogue record for this book is available from the British Library.

Every reasonable effort has been made to trace copyright-holders of
material reproduced in this book, but if any have been inadvertently
overlooked the publishers would be glad to hear from them.

Printed in the UK
10 9 8 7 6 5 4 3 2 1

FSC
www.fsc.org

MIX
Paper | Supporting
responsible forestry
FSC® C171272

In memory of James Lovelock, Polly Higgins and Professor Saleemul Huq. These three heroes have shone a bright light for so many and continue to inspire so many of us around the world as we face a difficult future.

Contents

Preface

The first Earth Day was organised in 1970 to demonstrate support for environmental protection. In 1972, the first United Nations Conference on the Human Environment, also known as the Stockholm Conference, held in Sweden, focused on air and water pollution, deforestation and the depletion of natural resources. It resulted in the adoption of the Stockholm Declaration, which recognised the fundamental right of all people to a healthy environment, and was attended by representatives of 113 countries. Scientists understood the planetary warming potential of greenhouse gases as far back as the nineteenth century. In the late 1980s, US scientists testified before Congress that carbon dioxide (CO_2) emissions needed to be curbed to safeguard the environment for global citizens and for future generations.

The Montreal Protocol, a global treaty signed in 1987, led to the banning of industrial substances, mainly chlorofluorocarbons (CFCs), that were causing a large-scale depletion of the protective ozone layer that sits in the stratosphere. The treaty was described by President Ronald Reagan the following year as, 'a model of co-operation. It is a product of the recognition and international consensus that ozone depletion is a global problem, both in terms of its causes and its effects. The protocol is the result of an extraordinary process of scientific study, negotiations among representatives of the business and environmental communities,

and international diplomacy. It is a monumental achievement.'
The Montreal Protocol led to the banning of ozone-depleting industrial refrigerants that was indeed a landmark moment in diplomacy. However, there are scientists today who say that it also injected an unhealthy amount of hubris into humanity's attitude towards clearing up the environmental disasters caused by human activities.

The success of the Montreal Protocol led to the United Nations Framework Convention on Climate Change (UNFCCC) borrowing the following ambitious line for its own treaty, stating that it 'bound member states to act in the interests of human safety, even in the face of scientific uncertainty'. The last bit about 'scientific uncertainty' is curious because, in recent years, it has emerged that oil companies, the main source of human emissions, had commissioned their own scientific reports accurately predicting the effects of the pollution they were causing, back in the 1970s. Skyrocketing profits from business-as-usual fossil-fuel production led them to bury their findings. These corporations then went further, casting doubt on the work of a growing number of independent scientists who confirmed the same findings that they themselves knew to be true. In 1988, the Intergovernmental Panel on Climate Change (IPCC), a scientific body involving hundreds of scientists from around the world, was set up by the United Nations Environment Programme (UNEP) and the World Meteorological Organisation (WMO) to provide policymakers with objective assessments and possible response options to the changing climate.

Scientists volunteer their time and expertise to assess the available scientific literature on climate-change. The comprehensive reports generated by the IPCC assess the physical scientific basis of climate, including impacts, vulnerability, adaptation and ways to mitigate greenhouse-gas emissions. The reports also include a 'Summary for Policymakers', containing summaries for informing on climate-related policies and actions. Although the IPCC has played an instrumental role in shaping international climate-change policies, including the UNFCCC and eventually the Paris Agreement, there has also been criticism regarding how reports can take years to compile and how the Summary for Policymakers is dumbed down to suit the interests of fossil-fuel corporations and fossil-fuel-producing nations.

The UNFCCC, created to address climate-change at the international level, opened for signature at the United Nations Conference on Environment and Development in Rio de Janeiro on 4 June 1992. One hundred and fifty-four countries and the European Union signed up to the convention on that day. As of February 2023, 197 countries have signed the declaration. Maurice Strong, Secretary General for the Environment, speaking at the Rio 'Earth Summit' in 1992, said:

> You come here as leaders of the nations of the Earth, and you are called upon to act as custodians of its future as a secure and hospitable home for present and future generations. I am deeply convinced that we are on a course that must change. There can be no doubt to anyone who goes around this planet and looks at the deterioration that has occurred in the last twenty years. A billion more people scrambling to make a living from a diminishing supply of resources in the south. We cannot be complacent. Unless the agreements reached here are accompanied by real commitments to significant change – change of course for the human species – we are, in my view, headed for a moment in the twenty-first century where the condition of our species may become terminal.
>
> What we need here are the political commitments that only you can make. You have the power. You have this unique opportunity and, most of all, you have the responsibility.

Over thirty years have passed, and carbon pollution continues to rise to record levels. Since the signing of the Paris Agreement in 2015 (two weeks after the death of Maurice Strong, quoted above), we have collectively (although not equally) emitted in excess of 250 billion metric tonnes (i.e., 1,000 kilograms per tonne) of heat-trapping gases into the atmosphere.

In 2022, it was announced at COP27 in Sharm El-Sheikh that global carbon emissions from burning fossil fuels had risen again by 1% to 37.5 billion tonnes. By 2030, they are estimated to be 16% higher than today. For some context, the first Conference of the Parties, COP1, took place in 1995 in Berlin. At that time, the Marshall Islands, fearing complete loss of their country to sea-level rise, expressed resentment at countries that sought to undermine the convention. The United States said they were committed to lowering carbon emissions to 1990 levels, and that they hoped an agreement could be on the table in 1997. Kuwait and Saudi Arabia 'expressed concern about the economic and social impacts of lowered emissions', with the latter adding that consideration of 'emission reductions impeded the progress of the Convention and overlooked possible *positive* effects of climate-change on agriculture'.

In 1997, the Kyoto Protocol, the first treaty to reduce fossil-fuels, was adopted but not ratified until 2005. In 1998, Al Gore signed the treaty on behalf of the US, but in Bonn in 2001, George W. Bush failed to ratify it. With the US responsible at the time for 25% of global emissions, this was a huge abdication of responsibility. Also in Bonn, the treaty was weakened to allow mega-polluters Japan, Canada and Australia to sign up (Australia finally signed in 2007). It was only when Russia signed up to the treaty in 2005 that the threshold of signatories, equal to 55% of global emissions, was crossed, allowing the treaty to become legally binding. The US remained outside of the treaty.

Emissions continued to rise all the way to the next milestone, COP15 in Copenhagen, where more attention was being given to the proceedings. The media pulling power of President Barack Obama instilled renewed expectations for a planet-saving deal, with the conference being dubbed 'Hopenhagen'. It is hard to comprehend what was really going on in Copenhagen, but it became apparent that the true 'climate nutters' at the COP were not the activists. They were, in fact, the world leaders and negotiators entrusted with getting us out of this mess. The Danish venue itself turned out to be too small, so many of the talks were conducted without official NGO observers present, adding to the murkiness of the proceedings. There are stories of Obama and Hillary Clinton bursting into meetings, surprising officials from developing nations (Brazil, South Africa, India and China) to apply pressure on them to sign up to the Copenhagen Accord, a non-binding agreement that vaguely aspired to limit global heating to 2°C. China and India blocked all moves to create a binding treaty to limit global heating to 1.5°C by reducing carbon emissions by 80% globally by 2050. What emerged from 'Nopenhagen' was a dumbed-down post-Kyoto agreement with no teeth, setting the scene for the kind of *ambition*-based deal-making that emerged from Paris.

This book covers my reporting from COP21 in Paris, a megaCOP of over 40,000 attendees, to COP27 in Sharm El-Sheikh, an unintended megaCOP of equal attendance. I have included many quotes from my interviews with scientists, activists, politicians and other key voices, as well as recorded speeches and presentations of established thinkers on this subject.

Since COPs began, the impacts of anthropogenic (man-made) global climate-change are being felt around the world, both directly and indirectly, via extreme seasonal weather patterns and a continual slew of recorded disaster footage posted across news and social-media channels. Actual events are replacing the warnings, strengthening the arguments of activists who are protesting across the world for humanity to change course. The new breed of climate activists are highly visible and can be seen blocking busy streets, storming the runways of private-aircraft owners or throwing soup at a van Gogh painting.

Whether we agree with their actions is immaterial because, in their efforts to express the warnings scientists have issued, only an appropriate response from society can diffuse their determination. Those we consider to be 'climate nutters' follow in the footsteps of suffragettes and civil rights movement activists who risked all to affect change. Many youths, parents, uncles, aunts, grandparents and other concerned citizens from across our society have reached the same conclusion: the future we are bringing on ourselves, and that has already arrived for many in vulnerable parts of the world, is looking so bad, it is worth doing whatever it takes to sound the alarm bell. The sound of the alarm bell is an attempt to change the powerful system of consumption.

But who is the alarm bell being sounded for? In the elapsed time of the eight COPs I have attended, a great deal has changed in the social and political space that shapes the conferences and yet nothing has achieved the original goal of lowering global carbon emissions. In conversations with COP veterans, scientists, politicians, activists and many others, we see the pathway emerging that will take us towards our uncertain future. For the rest of this decade and into the next, we will probably see the COPs' role in modern society change much more as climate impacts intersect with a wide range of issues, including global food production, supply chains, mass migrations, conflicts and a range of technical proposals to cool the planet.

Many in officialdom accept that we will exceed the warming boundaries scientists have warned against. New organisations are being formed looking at the options, or rather, *interventions*, available to us in these circumstances and how they might be applied. Real climate nutters are not just those people blocking the road or the suited individuals clapping in inadequate agreements at UN conferences. They also include the engineers spraying particles into the stratosphere, the social scientists saying we shouldn't, as well as the multitudes at the gate, confused as to how it ended up this way.

Introduction

Prior to the start of my investigations into what is going on with our climate, I was involved in a project to create a series of documentaries with leading British artists. Much of the series hinged on the young singer-songwriter Amy Winehouse, who had loosely agreed to be painted by the father of British Pop Art, Sir Peter Blake. The series was to be called *Icons of Art*, but after Amy's early death, the series had to be abandoned.

With my colleague and friend, Robin, I recorded a shorter documentary with the British artist Maggi Hambling, called *Dead or Alive*. We shot the film at the Tate Gallery, National Portrait Gallery, British Museum, by her sculpture of Oscar Wilde in Charing Cross, London, and in Maggi's studio in South London. We focused on Maggi's intimate relationship with her subjects as she recorded their portraits from both life and death. She worked beside her models as they lay cold in the morgue, drawing on memories and dreams, continuing her conversation with them in her head.

It was during this period of immersion in the existential theatre of Maggi's work that an unexpected subject came into view. The scale and scope of climate-change brought all other endeavours into question. I remember reading an interview in *The Sunday Times* with a scientist talking about the state of the Earth. The interview was with Professor James Lovelock about his soon-to-be-published book, *The Vanishing Face of Gaia*. As I read the words of the man in his early nineties, a steady stream of doubts about the balance of nature and humanity emerged and meshed with Lovelock's words, dominating my thoughts. Drawing the newspaper close, I read with undivided attention, and pre-ordered the book the next day. When *The Vanishing Face of Gaia* arrived in the post, I devoured it in a saddened state of comprehension. The glimpses between the cracks of my comfortable life revealed a troubling new story. Deep down, I sensed something was rising, as my world about me changed. Lovelock's long life scattered genius insights into the workings of our world and to the planets and stars beyond. When I interviewed him on the eve of his ninety-ninth birthday at his final home, close to Chesil Beach, Dorset, he displayed his trademark mischievous clarity. We discussed his own personal achievements of the past but also the future and what wonders he believed lay ahead.

Back in 2009, it was the eloquence and simplicity with which he conveyed the sensitivity of the Earth system to temperature changes, and how the activities of humans, mainly by burning fossil fuels, were causing a new rapid heating of the biosphere. His voice joined a wider chorus, warning that our civilisation could not adapt on any comparable scale to achieve anything like many of us experience today.

My instinct was to perceive this as an existential threat to myself, my friends and my family. Personally, I felt a sense of terror. I spoke to others about what I was thinking, but largely my concerns were ignored or explained away in non-satisfactory ways. My innately passive demeanour allowed this explaining away to happen, despite the inner turmoil that was stirring.

This was my initiation into becoming a climate nutter. I felt dismayed and at times overwhelmed. I gave the book to my friend Andy Worboys, a film and TV editor, who has since received multiple awards for his work. Andy read the book and agreed with my concerns. He ordered his own copy and, with an array of Post-its and other notes, morphed the tome into a three-dimensional sculpture. We then contacted Lovelock and met him in his hotel at Earl's Court. It was the morning after he had participated in a live conversation with his friend Sir Crispin Tickell at the Southbank Centre, promoting the same book. The hotel was in a back-street Victorian townhouse, the likes of which seem to be everywhere and nowhere across swathes of West London.

We rang the bell and a weary manager opened the door. He responded to our request for Jim by nodding us in and calling up to his room. Within seconds, a chirpy ninety-year-old scientist sprang from a shadowy staircase and beckoned us to follow. We bounded up three flights of stairs into Jim and his wife Sandy's hotel room and perched on the end of their bed for a discussion about our project. We had prospectively given it the title *A Hitchhiker's Guide to Gaia*. Jim loved the concept and agreed to participate.

The next time we met was in Cornwall at his remote home surrounded by woodlands, meadows, a large pond and remnants of the old Bodmin railway line. He took us on a rainy walk around the property. Jim's energy and agility were incredible. We struggled to follow him, sliding down wet banks and pushing through his overgrown wilderness. He would gesture with his arm towards some wild-growing woods densely packed with brambles and gorse, to declare that here Gaia was left alone to do whatever she pleased.

This whole experience with Lovelock was rewarding and insightful, but the project was ultimately swept away in the wake of the scandal that became widely known as Climategate. The fossil-fuel industry and their various paid mouthpieces in the mainstream media targeted real scientists. They hacked emails, lies were told, and in the remaining confusion, any mention of climate-change was cut from media platforms. We lost the best part of a decade in the UK for discussing, let alone tackling, the climate threat. The scientists were all proved innocent of the charges, but the damage had been done. Slurs and misrepresentation of the science continued in much of the highly trusted but malign mainstream media.

Andy went back to making films, but I decided I had to continue to find out if the risks posed by climate-change were as bad as Jim had made out. I started emailing scientists and requesting interviews. One early interviewee was Professor Peter Wadhams, based at the Department of Applied Mathematics in Cambridge. Peter was the foremost authority in the world on polar regions. In one interview, he recalled arriving at 10 Downing Street to a reception where, on seeing him, Prime Minister Margaret Thatcher had turned and yelled, 'Dennis, it's the ice man!'

Many Thatcher-era politicians of the same party formed the backbone of the climate-denial machine, not sharing her respect for science. I witnessed one such individual on the prime-time BBC programme *Newsnight* feigning outrage that his anti-science views were being countered by a real scientist, whom he declared to be an alarmist. One of Professor Wadhams's major contributions was to gather data from British and American nuclear submarines on exercise beneath the ice-covered Arctic Ocean. This data showed that the *volume* of the sea ice was thinning much faster than anyone had realised. This sea-ice decline is much more pronounced today and was declared in 2021 by the Climate Crisis Advisory Group, led by former UK Chief Scientific Advisor Sir David King, to be beyond its tipping point of irreversibility.

It was in 2012 when, with the help of an American, Gary Hauser, I arranged to interview Professor James Hansen, then director of NASA's Goddard Institute for Space Studies, during his visit to Vienna for the European Geophysical Union

(EGU) conference. Hansen is world famous for testifying to US Congress in the late eighties, stating the world must transition away from fossil fuels to avoid exacerbating global warming and causing significant risk to humanity. In the world of climate science, he is still a major figure, with his Indiana Jones-style hat, poignantly accurate climate science and stark critique of the political and industrial establishment in responding to what he recently called the 'Big Climate Short'. In the weeks before COP28, Hansen released new research. The subject of his email announcement was: 'How We Know that Global Warming Is Accelerating and that the Goal of the Paris Agreement is Dead'.

The weight of what I was learning, set against the complete silence in the mainstream media, created an itch of anxiety I could not rid myself of. The following year, I attended Al Gore's Climate Reality Training in Istanbul.

The Istanbul visit turned out to be interesting for various unforeseen reasons. On the flight over, a young man next to me asked if I was going to join the protests. In recent days, parts of the city had erupted into violent clashes with the authorities, as residents protested against the development of a small area called Gezi Park, next to Taksim Square. The wider issue that inflamed the sense of injustice was the obvious corruption of Turkey's elites, which appeared to include Prime Minister Erdoğan, a former mayor of the city. I told my fellow passenger that I was not heading to Istanbul for the protests. He replied that he had been in Cairo for the protests weeks earlier and had enjoyed the experience immensely. I considered this as I sipped my way through several in-flight Turkish red wines. Turkey is consistently one of the largest producers of wine in the world, with an incredibly long history of viticulture. I was only too pleased to indulge as my new friend drifted into a deep sleep.

From my hotel in the Sultanahmet district, I found it near impossible to find a taxi that would go near the protest area where Gore's training session was taking place, a short walk from Taksim Square. Eventually, I got there, and the training conference began. Events turned awry when Gore invited the Deputy Prime Minister of Turkey on stage to tell us what the country was achieving in its eco efforts. Immediately after he finished his slideshow, around sixty Turks in the audience stood up and chanted in unison the slogan, *'Heryer Taksim, heryer direniş,'* which translates as, 'Everywhere is Taksim, everywhere is resistance.' Adrenaline infused the mood as the protest entered the enclave. Security rushed Deputy Prime Minister Ali Babacan off the stage as the chants continued. Gore re-emerged with polite thanks for the official and restrained enthusiasm for the response of the audience.

That evening, I and a few others left the conference centre and joined the throng of protestors walking towards Taksim Square. The air was thick with tear gas, and the closer we got to the square, the more the noxious chemical irritated our eyes and breathing. My main observation was that the protestors were young, fashionable youths who could just as easily have been going to a pub or disco. They moved in their thousands towards the impact zone where police were waiting. We arrived at the edge of Taksim Square in a disorientating fog of gas-filled air and loud, rhythmic chanting of protest slogans. In what was already a chaotic scene, police started charging from both sides and firing more gas canisters directly at us. Our group of four split in two as we were absorbed into the huge crowds fleeing, being channelled down back streets. Above us, older people leaned out of windows, banging pots and pans, creating an abrasive din in solidarity with the protestors below. The atmosphere remained tense. It was now clear the police were using violence against citizens, as youths appeared in shocked anger, with torn clothes and cuts.

We waited in a narrow back street during a lull in the turmoil. As the banging of pots and pans became louder and more intense, an elderly lady opened her door and invited us in off the street. We declined, and as she shut the door, we heard an enormous pounding of feet growing louder. We ran, finding ourselves at the front of the charging crowd, and rounding the corner to a crossroads, we saw an equally large crowd running straight towards us and another crowd coming down on our right, towards the intersection. Only one road remained. We sprinted at full speed down the hill out onto a wide road at the bottom, hoping no ambush awaited us. We were about thirty seconds ahead of the police, who were racing towards us to seal off the junction. We charged across the road to a large hotel, piling through the security guards and metal detectors at the door and into the hotel bar.

By morning, the chaos had escalated, and the city was in turmoil. The climate conference was cancelled and we were advised to leave the area. Those who made it in headed straight for the bathrooms to wash their eyes clean of the irritating gas. Over successive days, the fighting grew worse and lives were lost. I thought of those young people I saw in the march towards the square and wondered how many had been brutally hurt during the protests.

As environmental protests are spreading around the world, with tactics becoming more extreme, the public are split on whether to agree with such actions. There are many who regard protest groups, such as Extinction Rebellion, Just Stop Oil or Insulate Britain, as nothing but fringe lunatics who should be locked up. In a complex social system where urgent change is needed but held back by a resilient status quo, we are all forced to consider the question: who are the real lunatics? As the natural world around us deteriorates at an accelerating pace, I have come to believe it is all of us.

Glossary

ADNOC	Abu Dhabi National Oil Company
AOSIS	Alliance of Small Island States
AR5	Assessment Report 5 of the IPCC
ARC	Australian Research Council
ASU	Arizona State University
CCAG	Climate Crisis Advisory Group
CCLS	Cambridge Climate Lecture Series
CCR	Centre for Climate Repair
CDM	Clean Development Mechanism
CFCs	Chlorofluorocarbons
COP	Conference of the Parties
EACOP	East African Crude Oil Pipeline
EGU	European Geophysical Union
GDP	Gross Domestic Product
GMACCC	Global Military Advisory Council on Climate Change
INDCs	Intended Nationally Determined Contributions
IPCC	Intergovernmental Panel on Climate Change
ISIL	Islamic State of Iraq and Levant
IWCA	International Wineries for Climate Action
KXL	Fourth pipeline of the Keystone Pipeline System, linking Alberta's oil production centres with ports and distribution points across Canada and the US
LDCs	Least Developed Countries

NDCs	Nationally Determined Contributions
NETs	Negative Emissions Technologies
NGO(s)	Non-governmental Organisation(s)
PIK	Potsdam Institute for Climate Impact Research
SDGs	Sustainable Development Goals
SIDS	Small Island Developing States
SRM	Solar-radiation modification
UKYCC	UK Youth Climate Coalition
UNEP	United Nations Environment Programme
UNFCCC	United Nations Framework Convention on Climate Change
WMO	World Meteorological Organisation
WTO	World Trade Organization
XR	Extinction Rebellion
YOUNGO	International Youth Climate Movement, the official children and youth constituency of the UNFCCC

1

COP21, Paris, 2015, Part 1

Filing through the queues at Saint Pancras Station in central London to get the Eurostar to Paris, I had mixed feelings of excitement and trepidation. COP21 was being billed as the means to get the world back on track to tackle the worsening climate problem. A deal between nations was on the table. The Executive Secretary of the United Nations Framework Convention on Climate Change, or UNFCCC, Christiana Figueres, a composed, fast-talking Costa Rican diplomat, was determined to bring the nations of the world to heel on this issue. Figueres had been traversing the globe, speaking with world leaders and their representatives, with rock stars, activists, religious leaders and everyone in between, in order to bring together the nations of the world at the twenty-first Conference of the Parties (COP21) to sign a global agreement to bring an end to fossil-fuel consumption.

The momentum in the run-up to COP21 was severely hindered when 130 people were murdered in the centre of Paris, ninety of whom were inside the Bataclan theatre, leading to the bloodshed being dubbed 'The Bataclan Massacre'. The Islamic State of Iraq and Levant (ISIL) claimed responsibility for the attacks in retaliation for French airstrikes on Islamic State targets in Syria and Iraq. A staggering 413 injured accompanied the death toll, about one hundred of whom were critical. President François Hollande immediately declared a state of national emergency. In Paris, the shock and trauma of what had happened remained constant, as heavily armed security forces were widely deployed, not only to protect citizens but also to signal clearly to the world that COP21 would go ahead as planned. With an expected attendance of over 40,000 people and 150 world leaders, the largest ever in attendance at a UN event on the same day, there was a lot at stake.

Prior to the Bataclan attacks, interest in COP21 Paris had been rising. Al Gore's 2006 documentary *An Inconvenient Truth* had created widespread awareness of the climate threat, but, up against the fog of fossil-fuel industry disinformation, the memory of what it was about had faded.

Paris was looming, and the media was perking up with curiosity. Were we now on the cusp of a change that would see world leaders do a volte-face and sign a deal into being to limit the use of fossil fuels worldwide? This question was especially pertinent in rich nations that had benefited from industrialisation fuelled by coal, oil and gas.

It was now time to cut back, reduce our impact and decarbonise our economies. The poorest developing nations needed to raise their standard of living in the face of worsening impacts, while also building resilience to survive the future. To achieve this, they would have to leapfrog all the dirty fossil energy, transitioning straight to low-carbon energy production. Christiana Figueres spoke to every country's delegation on Earth, asking them to sign up to a deal promising a worldwide phasing out of fossil-fuel production and use. Transitioning away from coal, oil and gas would set us on the right path to stabilise the climate and leave a survivable planet for future generations. Not to speak of the ecological and social benefits of no longer consuming as much rubbish as we do in the wealthier parts of the world.

Despite the good intentions, the reality of these conferences is that every country attending can send as many official negotiators as they choose to the COP negotiating process. The average is twenty per nation, which adds up to around 4,000 negotiators, meeting for two weeks. Working as the science lead for the UK's Foreign Office in Paris, Sir David King recalls, 'How do you negotiate in that sort of atmosphere? I had explained to the heads of governments I was working for that we need to have bilateral action. That is why I visited ninety-six countries on official visits in the run-up to Paris. Those bilateral negotiations were, for me, the only way of breaking the logjam.'

Each nation produced its own 'Intended Nationally Determined Contributions', or INDCs, for reducing fossil-fuel emissions. After COP21, with the Paris deal signed, these INDCs would no longer be *intended*. They would be on the path to implementation, and therefore just NDCs.

There is a classic weakness to this whole effort that cannot be ignored. The Paris Agreement could not have any strength to punish countries if they did not follow through on their NDCs. The proposed deal was not legally binding. It was voluntary. If there was good reason to be cautious, the precedent of previous summits provided it. Not much had changed in the actions of high-polluting nation states and high-emitting corporations throughout the period since Rio in 1992, or even since COP1 Berlin in 1995. For example, was the United States Senate about to back a climate treaty that limited their excesses? As George Bush Senior said at the time of the Rio Earth Summit, 'The American way of life is not up for negotiation. Period.' Two decades on from the first COP, was COP21 Paris to be *the* climate summit to save humanity with an agreement that was non-binding?

Exiting the tunnel onto French soil, the landscape appeared stark and monochrome beneath an overcast sky, the dourness punctuated by the aroma of my scorching hot, bitter coffee and the better-than-expected croissant. My phone was buzzing with messages from an eccentric American COP veteran called Scott, whom I had been liaising with and had agreed to collaborate with. The messages were varied: 'We are going to get food.' 'When are you arriving?' 'STAY AWAY FROM THE PLACE DE LA REPUBLIQUE... SHOWDOWN!! WILL TURN NASTY!!'

I had to scan the newsfeeds to see what was going on. It quickly emerged that people labelled 'environmental protesters' were clashing violently with riot police, in what would end up leading to 200 arrests. Scott was right; it was best to stay away. In the run-up to COP, I had been getting an increasing flow of press emails from the UNFCCC media list, which, when I had the time to sift through the general detritus, could lead to an occasional announcement of interest. That morning, I had received a media invitation to attend a press-conference given by Canada's new Minister of Environment and Climate Change, Catherine McKenna, at the Canadian Embassy that same afternoon. I calculated that if I jumped on the Metro, straight from Gare du Nord railway station, I could make it just in time.

Canada's Athabasca 'tar sands' are widely reported by environmental NGOs, climate scientists and established media alike as the most destructive oil-extraction site on the planet, covering an area of land larger than England. It is an ongoing man-made ecological disaster, not only poisoning the land but destroying the health and communities of indigenous peoples since oil production began at the site in 1968. At the current rate of extraction in 2023, 2.8 million barrels per day are being produced in Northern Alberta. The then director of NASA's Goddard Institute for Space Studies, Professor James Hansen, had said to *The Guardian* newspaper in 2013 that further exploitation of Canada's tar sands would be 'game over for the climate'.

Environmentalists heavily criticised the outgoing administration, led by Prime Minister Steven Harper, for their glib denouncement of environmental concerns. Emblematic of the struggle to stop the export of this dangerous fuel is the Keystone Pipeline System, linking Alberta's production facilities with various ports and other facilities across Canada and the United States. The Keystone Pipeline had progressed to Phase 3 in 2013, during President Obama's term in office. Keystone ran all the way from Alberta in Canada to Port Arthur, Texas, in the Gulf of Mexico, despite being bitterly opposed at every stage.

Running up to COP21, the Obama administration vetoed the next phase of the project to build a fourth pipeline known as Keystone XL (KXL), citing environmental and economic concerns. There was a slight glimmer of hope that the new, fresh-faced Trudeau administration would arrive in Paris announcing that the Keystone development was dead and that they would scale the tar sands down.

As the Eurostar train pulled in to Gare du Nord, I readied myself by the train door. My suitcase was absolutely huge, bulging with equipment, feeling as though it weighed twice my body weight. Convinced of every item's necessity, I didn't begrudge the strain of lowering it onto the platform. From there, I marched off through the ticket barrier and into the Metro station. I quickly realised how heavy my suitcase was as it came crashing down the stairway onto the busy Metro platform, landing on top of me.

The familiar aroma of subterranean Paris permeated my senses: a blend of scorched woodiness fused with the industrial grind of the train tracks. By the time I arrived at Saint-Philippe-du-Roule and negotiated the case up the stairs to the street, I was sweating and panting heavily. A two-minute jog further and I was standing dishevelled at the embassy entrance, filing in to get through security, with very little time to get a spot and set up my camera. The briefing room was compact, with the obligatory national flags flanking two high-backed leather chairs with a black-cloth-covered table forming a buffer between stage and pit. Between the flags, in faint lettering, the words 'Together, For The Climate' pinned the occasion to the cause. Large cameras were mounted, and journalists occupied the several rows of seats, which gave way to a tightly packed mob at the back. I too found a space a little way back but with a clear cross-view to the speaker's podium. The cameraman directly ahead of me glowered in case I was thinking of encroaching on his patch.

A press officer came out and introduced the ambassador who walked out, and introduced McKenna. The minister was wearing a black dress and a blazer in a striking hue of Canadian red. Within moments, Catherine McKenna beamed at the room and delivered what was to be the Canadian meme for COP21: 'Canada is back!' In fact, McKenna seemed nervous and her performance was not convincing. Each answer to questions posed seemed to be a jaded recollection of a poorly written press release: 'Canada is back! We are coming to the table with a lot of optimism and a lot of ambition. I am confident that we will leave the table with a way forward towards genuine progress on climate-change. So, we need an agreement that is going to be legally binding. In particular, we need the agreement to be binding in terms of each country committing to a target. To each country committing to transparency and accountability, and provisions to improving on the target every five years.'

So far, so good, but after this, the script backed away from the 'need, the commitments and the accountability' towards a mad moment of honesty on McKenna's part: 'There will be some parts that, you know, to reflect the reality that certain countries have concerns with all aspects being binding. Maybe not all aspects will be binding but it's just really important that we have everyone at the table and come up with an agreement that's really critical.'

It is important to note that the weakening of the Paris Agreement to ensure that the commitments were not binding was because of the domestic policy in the US. President Obama could not sign up to the Paris Agreement if Congress then prohibited his administration from making the policy decisions required to fulfil the obligations. The wider problem with this is that the fossil-fuel lobby in the US is intertwined and emulated in other nations, taking cover from the US position. Canada did not need to close down the tar sands because the US was leading by example, albeit in the wrong direction. This was the same issue in other developed countries and blocs around the world, whether they were Australia, the EU, Britain, or anyone else.

As the questions homed in on the tar sands in Alberta, McKenna had the following to say: 'The oil sands are important to the Canadian economy. It's important to the Alberta economy, but we're very encouraged that the Government of Alberta recognise Alberta needs to be doing its fair share in terms of tackling climate-change. So they've come up with a number of different policies, including putting a cap on emissions from the oil sands that will make, you know, will make a difference!'

When it came to taking ambitious climate action, like the US, the Canadian way of life was also not up for negotiation. Inside the COP, I would get my chance to interview the Alberta Environment Minister, Shannon Philips, to ask her directly about the future of the tar sands. This was day zero of the conference and my optimism that the world would come together to make the structural changes necessary to rapidly reduce emissions was already shaken. I packed up and ran out of the embassy, heading directly back to Gare du Nord. Scott had reserved rooms for a few of us in a hostel called St Christopher's Inn, or during the COP, dubbed Place2Be, assuring me it was an insider's venue for the event.

The games begin

I dumped my stuff at the Place2Be reception, which looked more like a disco bar than any sort of hotel. Without time to think, I ran back into the station and took the free train to the COP21 venue at Le Bourget, Paris's second airport. Scott, a veteran of these conferences, had advised me to get my registration and ID badge sorted out now, before the crowds arrived, because the following morning, the desks would be in chaos with the inflow of arrivals.

When I returned to Place2Be, it was after 7.00 p.m. on the Sunday evening ahead of the opening of COP21, and although I had intended to do some prep work ahead of the following day, I changed my mind when I bumped into Scott in the bar area. As he greeted me, I noticed he was wearing a loose woollen knitted sweater of multicoloured stripes. He was very thin, almost gaunt, with thinning hair and a shifting smirk, underpinned by a hoarse New York drawl – he was amiable but intense. This relaxed version of him in the hippy Day-Glo woolly stripes would, in twenty-four hours, be switched into the razor-edged, at times maniacal, character for the duration of the COP.

I bought a beer and spoke to Scott about his plans for the next two weeks in Paris. Scott had a natural intensity that he cultivated into a personal theatre. He explained his plans for a 'Climate TV Show' within the conference: '[It] is a new concept for a TV show at the COP. It's a chat show with expert commentators who are going to tell the public what is really happening at the UN. Not the bullshit the mainstream media wants you to believe.'

This was the first I had heard of the daily COP Climate Show and it sounded intriguing. In 2015, very few media outlets published important articles about climate-change. In the UK, the *Guardian* and *Independent* were the only newspapers that did so consistently. The forces of climate denial, who had produced the fake Climategate scandal six years previously, in 2009, had left a legacy of mainstream media blackout. The result was that the UK had virtually no significant climate reporting to inform the public. Some extreme events that directly impacted people were reported in the news, like flooding or named storms, but the connection to climate science was consistently omitted. Fossil-fuel lobbyists in the UK, including corrupt lords and MPs, worked hard to ensure the public were never properly briefed. There was much money to be made, and there still is, in casting doubt on the claims of climate scientists and journalists.

Scott was planning a wider documentary and had asked me for help capturing footage. This I could not commit to as I had my own workload, but I had introduced him to Mike, a contact with decades of experience as a TV cameraman and sound engineer. Mike was keen to get more first-hand knowledge on climate issues, and what better opportunity than to be on the ground in Paris. Mike and his partner Lizzie arrived in Paris to assist Scott for at least part of the conference. It was getting late and the Place2Be was filling up with climate revellers. The music volume was rising and hardening into a speedy rhythmic beat. I said goodnight to Scott and headed to bed.

Into the COP bubble
The combination of beer and earplugs meant I had a fairly decent night's sleep, despite the pumping techno music until 2.00 a.m. I rose early, scrambling to gather my senses and my equipment. With a bag loaded with two film cameras, two tripods, mics and a laptop, I headed out of the door. The dress

code was business casual: shirt, jacket and trousers, with smart sneakers for comfort.

Downstairs at the breakfast buffet, other COP-badged guests were gathering at tables, and I grabbed the constitutional croissant, black coffee and orange juice, as well as a seat. Glancing around, I noticed Vandana Shiva, the Indian activist and academic, as well as others who I recognised from events in London. It was 7.00 a.m. and more people were filing in. Keen to stay ahead of the throng, I ate my breakfast at speed and headed for the street.

The journey to the COP was easy from Place2Be – Scott had chosen well. It took a couple of minutes to walk to Gare du Nord and catch a designated COP train, taking us a few stops to where the shuttle buses, working on rotation, were ferrying the thousands of delegates to the venue, Le Bourget. A COP delegate's ID badge entitles you to various other perks, such as free city-wide travel on public transport and a branded water bottle for refilling around the conference. About 40,000 people visited COP21, so that is a lot of people utilising the transport system. It is also a lot of extra people looking for accommodation, places to eat out and bars to drink in. That's the upside for the host city.

It was now getting on for 8.00 a.m. as the bus pulled up outside the entrance to Le Bourget. The crowds were thick, with more people appearing from every direction, but the flow was fast-moving. I noted to myself that I should start arriving at the gate at 7.30 a.m. to avoid the crush. The 'COP Village' is a vast complex. There is a main concourse that if you just walk it from end to end, it takes about twenty minutes. Running off it are several halls to accommodate various groups of delegates. With a media badge, the areas that I could not access were the negotiations. That is the domain of the Observer badge holders. Observers can go in and witness what is said in the meetings. Some of these observers will be legal advisers, interpreters, analysts, academics, lobbyists and so on. From what I have been told repeatedly, unless you are being paid to be in there and have an actual role or specific interest, it is the negotiations that are best avoided. Representatives spend hours arguing over word changes in the documents, attempting to skew meaning in favour of their nation's or organisation's interests.

The United States is one nation known for playing hardball in the negotiations, using its economic and political muscle to influence outcomes. As outsiders, we might hope that the collective good of the planet is the primary goal for every negotiator, fighting through the lens of their national interest. I asked climate law expert Professor Dan Bodansky, from Arizona State University, how, from his experience, the negotiating team would strike a balance between 'America First' and 'Planet First'. He told me:

> Well, I would say for every country that is here, they're trying to advance their national interests, so it would not be limited to the US. To a significant degree, though, there shouldn't be a disjunction between America first and planet first, because we have a stake in the planet surviving. The fact of the matter is, climate-change affects different countries differentially. Countries have different economic interests in addressing climate-change. For countries that highly depend on fossil fuels, it's more difficult for them to decarbonise, since fossil fuels are a significant source of income.
>
> So one thing that makes the COP so complicated is that everybody's national interests are not perfectly aligned. If you are a country representative, your job is to represent your country's interest as your political leaders see your country's interest. You're not acting as an ambassador for the planet.

Hopefully, and I think this is the case for many countries, countries have an interest in addressing climate-change and are trying to address climate-change in a way that's compatible with their interests. It is also the case that what you can do internationally is very much dependent on what is politically possible domestically. So if you are just saying 'planet first', you can make all sorts of statements here, but you will not be able to actually come back home and implement them because you have to get things through your domestic political process.

Dan's succinct overview of the negotiating parameters is logically pragmatic but, paradoxically, is also what can make the lack of progress at the UN level so frustrating.

Back inside the Blue Zone, we have the main conference halls where the celebrities and VIPs come to give speeches to energise the momentum for change. Prominent VIPs of all stripes give speeches, including politicians, diplomats and celebrities, quoting the warnings issued by scientists and the imperative for the negotiations to be instrumental in bringing carbon emissions down to zero. Revert to Maurice Strong's speech at the 1992 Rio Earth Summit and you have a template that will take you all the way from Rio 1992 to Paris 2015 and beyond, give or take a few edits. The speeches have continued in earnest, carbon pollution has continued to rise steeply, and every COP has failed to deliver on its objective to bend the emissions curve downwards.

Away from the main conference room, we have the hall where nation states and blocs, corporations, and large NGOs erect impressive pavilions to celebrate their culture, innovations and technology. In garish displays, dreams are cultivated for a future that is increasingly divergent from the reality we are speeding towards. Then there is a hall for smaller NGOs, academic institutions, faith groups and climate-related organisations. In Paris, this was among the most buzzy settings. Seating areas spread around the place: coffee bars, banks of computer workstations, superfast broadband, people hanging out, working, writing, interviewing or hosting meetings. After this, we eventually arrive at the media centre. This is a large and chaotic area where every news outlet has reserved space, with news-broadcasting areas, more long rows of workstations, booths where people can work in private, meeting rooms and three large, seated press-conference rooms.

It was day one and world leaders were arriving to open the conference. In the media centre, a torrent of hundreds of media people flapped about, contributing to the almighty din, and amid this chaos, news reporters were being filmed recording their segments to be livestreamed or sent back to wherever. I set up my camera and checked that I had everything. I then plugged in my laptop, connected to the wi-fi and scanned the UNFCCC press emails flooding my inbox. I quickly followed up by emailing a few potential interviewees that I knew would be at the COP. One was Dr Saleemul Huq from Bangladesh. I had interviewed him recently in Istanbul at the Islamic Declaration on Climate Change, as well as once before in London. Saleem is a lucid thinker and clear speaker. Bangladesh is one of the countries most vulnerable to climate-change. In Paris, they also had the largest number of press reporters who attended from start to finish, informing their people of the goings-on. Saleem is an advisor to the Bangladeshi delegation and a representative of numerous organisations, including being a founding director of the Independent University, Bangladesh.

I also emailed the then leader of the British Green Party, Caroline Lucas MP, to ask if she was coming to Paris. I didn't know Caroline personally, but she has remained an incredibly popular British politician, demonstrating integrity and a clear comprehension of the broader environmental challenges we face. The high level of chaos was energising and daunting at the same time. I started feeling like I should be more pumped with caffeine. That I should be more excited about the arrival of Prince Charles or UK Prime Minister David Cameron, or of President Obama, Prime Minister Modi or China's Jinping, and the list goes on. That was why the glossy media were here, inflated with optimism, to report the mainstream message that the power of the global political elites was going to save us. This gathering in Paris was pitched as that significant moment of change in our collective fate.

With laptop open and one camera up on the tripod, mics ready, and my mind fixated on coffee, I leant over to the woman opposite and said, 'Do you know where the nearest coffee place is?'

She smiled and said, in a German accent, 'You can buy coffee around that corner, but if you want a tip, the German pavilion has a free coffee bar. It is the best!'

'Where is that?' I asked.

'It's not far, actually. You take the exit over there and cross straight over. The German pavilion is along the dividing wall in the same hall.'

This was the kind of tip I needed. Leaving my gear, I jogged out of the door and across the concourse. This was my first time inside the pavilion hall, and the scale of it was quite arresting. This really was where countries got to tell their best eco-fantasies. Enormous depictions of solar farms, wind turbines on the land and at sea, and charts projecting into the future, showing carbon emissions coming down. I would revisit this area when I had time and take in the fullness of the greenwash.

A very comfortable coffee bar comprised about fifty per cent of the German stand, and the rest was presentation areas and meeting rooms. I ordered two double espressos before commandeering a stool to sip them both into oblivion. Feeling a little racier than before, I legged it back to the media centre to find Scott discussing logistics for his Climate Show events with Mike and Lizzie. I led them inside and found a place for them to work. Scott was wearing a beige suit and dark-rimmed glasses. His receding hair was tinted an autumnal russet colour. His pallid face and thin lips were concentrated. I realised now that this was Scott in official COP mode. He addressed us: 'The first show will start at 11.00 a.m., that is, in ninety minutes. Being the first one, it will be low key. Things are going to get significantly livelier in the next few days. This opening show will get me in the mood.'

Scott was appropriating the press-conference room, with all its professional TV-standard facilities, as a recording studio and, although not strictly as the facility was intended, presenting it as a *show* with a live audience. The size of the audience would depend on the fame of the speakers. Press releases were released through the UN media-centre facilities and the final recordings of the show were available to take away on hard disc. Scott had realised at previous COPs that this press-conference facility was a valuable currency because most people who came here wanted to be seen doing something important in front of an official UN or COP logo.

To get access to the press-conference rooms to host a thirty-minute session, you needed to be either a registered government representative or a member of an admitted observer organisation. Media people were free to attend and record in any medium, but we could not book our own press-conference slots. Scott had found a couple of friendly NGOs that pretty much donated their slot-booking capabilities to him. In return, he acknowledged them at the start of the session. In this way, he branded his session 'The Climate Show' and presented it as a talk show with invited guests.

'So, who is your guest today?' I asked.

'It's a Buddhist foundation that has gifted me their slots. I promised them the first show,' he told me.

When I asked who else he had lined up, he said that he had persuaded Dr James Hansen to come to Paris and participate in one of the Climate Shows, to make a statement about the COP process. This was impressive. Hansen had never previously attended a COP, so this was a major coup for Scott. I was impressed. 'Well done, Scott!' I said. 'Who else?'

He was now in his stride. 'Quite a few, actually. I've asked Dr Hansen to do a couple of sessions, but I just got confirmation that Archbishop Rowan Williams will be here for one day and will participate. I asked my friend Archbishop Kykotis to sit on the panel too.'

Scott looked at his watch, then turned to Mike, saying, 'Mike and Lizzie, you *are* good to record extra clips of the Climate Show in the press-conference room, aren't you?'

'Yes, that sounds very straightforward,' Mike answered. 'Do you want me to film you walking towards the room or sitting down? Some extra cutaways?'

Scott loved this idea. It was more pro than he could have wished for. 'Oh, yeah! That sounds perfect,' he replied. 'Wow, yeah!' He was pleased.

I was curious to see Scott's Climate Show, but I also wanted to find my own narrative flow as to what was the purpose for myself and all of us being here. I decided I would sit through the first half of the session near the back of the room with my mounted camera and then quietly slip out.

Finding a few spare minutes, I walked out to the main concourse to see if there was anything in the bustling crowd that was worth observing. My gaze was drawn to a fast-moving huddle of people edging down the central area like a wave, sweeping people up in its path. The momentum was being amplified by curiosity, drawing more people into it, forming a magnetic pull. As it passed me, I found myself on tiptoes peering in, fighting back the urge to fall into the throng. In the centre, I saw the grinning face of Ban Ki-moon, Secretary-General of the United Nations. Like many bureaucrats, if I saw him in the supermarket, I would not have a clue that he was anyone of significance, but in this place, officials are elevated to hero status. That said, at COP21, Ban was playing second fiddle to Christiana Figueres, who appeared every few minutes on random screens around the place, dressed in a power suit, looking on top of her game.

I wandered into the area of the stalls of NGOs and academic institutions. In the first aisle, I saw Professor Kevin Anderson, deputy director of the Tyndall Centre at the University of Manchester, walking towards me. I held up a hand, and he stopped to see who this person was that he didn't recognise.

'Excuse me, Kevin, my name is Nick Breeze,' I introduced myself. 'Do you have time to record an interview about what you think the outcome of COP21 will be?'

He digested the information, eyed my badge, and replied in his rapid-fire way, 'I can tell you now if you like? A lot of hot air and a lot of burned carbon but no meaningful action.' Then he added with a smile, 'I am sitting down over there, on that stand, for most of the day. Do you want to come and find me when you are ready?'

'Yes, thanks. I have another commitment in about half an hour, but after that, I will come and find you. Is a filmed interview okay?'

'No problem at all. See you later.'

And off he strode.

Walking along a little further, I noticed Saleemul Huq sitting on a low stool, wearing a microphone and being interviewed. Probably not a great time to approach him. I took out my phone and checked my emails. He had replied already:

Hi Nick,

I am in Hall 4 and will be here most of today. Just come and find me.

Best,
Saleem

With material evidence of his whereabouts, I just had to get Scott's show over with, then return here with my camera. I headed back to my workstation. Scott was standing with Mike, who had his enormous camera rigged up. Lizzie was holding the tripod bag, and they were chatting about logistics. I saw some people in robes approaching. They were clearly Scott's guests for his show. They all met and exchanged greetings, followed by him turning towards us, signalling to get ready to record them approaching the press-conference room.

Scott and his entourage then entered the press-conference room. The timing is always tight, and this is the ungracious part of the proceedings. Every press-conference session naturally runs over time, but is ruthlessly driven out by the succeeding one. Scott was not about to let this nondescript press-conference in process cut into his time. He approached the stage a few minutes early, interrupting the speakers, shouting, 'Two minutes, please, folks. We are waiting!' Those seated on the panel displayed obvious displeasure but, to be fair to Scott, the room could likely hold 150 people and currently had six achingly bored souls enduring a turgid presentation. He started unpacking his laptop as they wrapped up their session.

The panel was all seated and Mike was set up in the front row, supplementing the two huge cameras with operators midway down each side of the room. There were an additional two at the back, all linked to a control suite in the corner for live editing. It was time to get comfortable and enjoy the show.

This session had attracted about two more audience members than the previous one. With the world leaders entering the COP Village, it was unlikely that people would attend this now. I sat towards the back and had my camera set up on the tripod, ready to go. The session started with Scott playing some strange music, followed by him welcoming people to the show. The optics were surreal. He was in the centre as the host, looking as pale as a white marble statue, flanked by two robed figures on each side. The Buddhists, by contrast, just looked Buddhist, but Scott conjured a façade of outward sanctity. I got the format and could see how it worked. He would introduce them and let them speak for a few minutes each, before occasionally cutting in with a statement or question.

As they spoke, I scrolled through my phone and noted down some questions for Professor Anderson. I usually aimed for between six and ten questions in a typical interview, allowing time for digressions and expansions on points. Once I had some questions down, I looked up, noting Scott introducing the next speaker by name. While they were speaking, I grabbed my tripod and bag and, crouching down, ran for the exit.

Outside, I walked at speed back towards the hall with all the NGOs. Kevin stood chatting with a young woman who looked like a student. There was a joviality to the scene, and he appeared to be a pleasant character. He recognised me and cut the conversation short, taking the person's card and waving it in the air, indicating that it was in safe hands. He was wearing a blue-and-red-checked shirt, and he seemed very much at ease. I set the shot up with him against a bright yellow exhibition booth in the background that offered a colour contrast and framed him very well.

'Do you want to preview the questions?' I asked.

'Oh, no thanks,' he told me. 'I quite like having to respond to things and trying to think in real time.'

Although many interviewees are happy to proceed without seeing the questions, others much prefer previewing them. The purpose of this interview was to gauge what Kevin, whose expertise is in climate and energy policy, actually thought could be achieved at this conference. I put on his lapel microphone, hit the record button and asked what he thought was behind the apparent optimism that the many thousands of people gathered here were feeling.

'I'm not sure that everyone that comes to these events really considers that climate-change is a pivotal issue,' he began. 'They're probably not that deeply involved in some discussions about, say, the mitigation agenda on 2°C, or even the Loss and Damage issue related to impacts.'

My next question was whether he thought reducing carbon emissions and economic growth were compatible. This question can also be translated to ask whether we in wealthier nations can continue to consume as much as we do, while transitioning to a cleaner energy system that powers our consumption. He gave a detailed and helpful answer:

If we are talking about the carbon budgets for a 2°C temperature rise, and we're talking about the wealthier parts of the world, then I would say, not in the short term. The reason is, if we are serious about climate-change, we know the carbon budgets that we have from the IPCC. If we want a reasonable chance of 2°C, then we cannot carry on doing what we're doing if the poorer parts of the world are to have reasonable access to that carbon as well. So in the short term, the only thing that is really going to keep emissions down are radical reductions in energy demand by those of us who consume a lot of energy. They typically live in the wealthier parts of the world, but not all. There are certainly many people in the UK who live in very poor-quality housing, have relatively low levels of consumption, don't drive or fly, or fly very little, and their emissions will be relatively low. So for those people, they're not part of the problem, and for them, I do not want to see their emissions go down.

Many people in the poorer parts of the world have incredibly low emissions, and we know that their welfare, their quality of life, is linked to their access to energy. In the short to medium term, that access to energy is primarily going to be fossil fuels. As much renewables as possible, but it's significantly going to be a chunk of fossil fuels in there. That means their emissions will go up, and that is a good thing because that means their quality of life is improving.

That means yours, mine, and people like us have to make even more significant reductions in our quality of life. So when we look at this growth issue, it is about growth for whom? If we think about what is important in a typical Western society, and perhaps more widely now, what do we aspire to?

Consuming more energy, consuming more goods, living in larger houses, bigger cars, flying more, and more exotic holidays, going business class. Everything that we aspire to, everything that says something about our role, our position in society, is about more carbon dioxide emissions. So progress and success are measured in dimensions of our society that mean more carbon emissions.

If we're serious about a 2°C carbon budget, you cannot measure a society like that. You start to say, well, these are no longer appropriate metrics in the short to medium term for dealing with climate-change. Once we've got a low-carbon energy supply in place, which will take us twenty or thirty years, then we can go back to living the sorts of lives we want to today, as long as it fits the broader sustainability criteria. From a carbon perspective, we need to get the carbon out of the system, and that requires us to reduce our energy demand in the short term.

What struck me here is that no one in the mainstream is having this discussion about cutting back on consumption or limiting economic growth. The push is consistently for the opposite, or at the most, fantasies based on so-called 'green growth'. It feels like the public are just participants in a system of destruction and cannot see a way out. It is not just the so-called elites, but also the middle classes around the world. Our societal values and aspirations have evolved to be intrinsically based on ever-rising carbon dioxide emissions.

Resetting our values begins with questioning what we are going to aspire to now and how we communicate those values. Wealthy Europeans, Japanese, Australians and North Americans have been responsible for the bulk of the historic emissions from our patterns of consumption, but now there are also the burgeoning middle classes throughout the world who are living more carbon-intensive lifestyles. The need for collective change is set against the terrible fact that most of these people around the world are not even engaged in the issue anyway. I asked Kevin if we were not asking too much, too late, for the momentous shift required to happen. He answered:

It is a huge ask. We should have done something about this twenty-five years ago. We had the first IPCC report in 1990. With the Rio Earth Summit in 1992, we've had a quarter of a century of abject failure on climate-change. Let's be absolutely blunt about that. The carbon emissions this year will be sixty per cent higher than they were when we first pretended to care about climate-change in 1990. We have squandered our carbon budget for 2°C and therefore we are left with the difficult situation we are in today. It's tempting to say, tough, that's the position we're in if you want to hold to 2°C. Remember, that is a target that civil society, that the political framework, has come up with, and 2°C will not be safe for many people around the globe. It is not an appropriate target if you're poor and live in a climatically vulnerable part of the southern hemisphere, but that is the sort of target that we've collectively come up with. If we want to hold that target, then we should have done something earlier. We haven't, and we now face these incredibly tough decisions. That's where we are.

At the time of this interview, the temperature of the planet was estimated to have risen by a mean average of 1°C globally since the beginning of the industrial revolution. The first signs of real climate disruption, even in so-called climatically stable nations such as the UK, were directly in the line of sight of the public and the mainstream media. The floods that engulfed the UK during COP21 highlighted how vulnerable communities were to the new, advancing era of extreme weather. Yet here we were talking about a target of *double the current warming* to prevent global climate disaster. Even hearing politicians here talking about 2°C as a *target* was worrying. Should we even call it a target?

I've never liked the language of target or goal. We don't have a target not to murder, or a goal not to murder. We have it as an obligation or a duty in our society. I think it's completely inappropriate that we see 2°C, which will mean that many people will live much more impoverished lives, and many people will die as a consequence, as a goal or a target? It is a duty; it is an obligation. We've known everything we need to know to do something about it for two or three decades. Let's be blunt about this: 2°C was decided by rich, relatively wealthy, typically white men in the northern hemisphere as being the appropriate threshold between acceptable and dangerous climate-change. Then there are many people who will be poor and living in the southern hemisphere, who are typically non-white, very low emitters of carbon dioxide, that will have already been impacted by climate-change. This is going to be on top of the already very challenging lives that they have, just trying to eke a living out in often difficult circumstances, anyway. So for them, it is going to be somewhere between dangerous and deadly. This is not a safe threshold.

Dangers of looking the other way

These points covered by Kevin struck me as the dominant news story we should be collectively setting aside our differences to tackle. Climate-science models provided simulations as to what mean average global temperature rises we would get depending on how much of various greenhouse gases we pump into the atmosphere. Before COP21, these forecasts pointed to a hellish 'business as usual' outcome if nothing was done.

Prior to coming to Paris, every nation had submitted what are called Intended Nationally Determined Contributions (INDCs) that outline what actions they would take after 2020 to reduce emissions, in line with staying below the 2°C limit on mean global heating but aiming to stay as close as possible to 1.5°C and then aiming for net-zero emissions by 2050.

The Potsdam Institute for Climate Impact Research (PIK) published a study just before the Paris COP that calculated the total sum of all the emissions reductions in the INDCs, to see where we would be if all the pledges were achieved. They found there would be nearly double the emissions that scientists had stated would be the danger threshold.

But even this oversimplifies the Earth system to a huge degree. The other issue scientists were strongly warning about were the nine identified climatic tipping points, or 'positive feedbacks', within the Earth system. Once any of these is crossed, these heating processes become self-sustaining and can set off the others. How they interact with each other is not well understood, but they are expected to significantly accelerate climate chaos. There are signs today that we have either just crossed, or are dangerously close to crossing, one or more tipping points. Staying with the 2°C *target* as a limit for global warming, I asked Kevin what he thought about the danger of crossing tipping points with temperatures that high.

There are many scientists already who think that a 2°C temperature rise, if we head towards that level, will be too high. We will kick into train a series of what are called positive feedbacks, or tipping points. They're not very positive. They're bad for us. They make the situation worse. Across the scientific community, there's a lot of uncertainty about exactly when those positive feedbacks will kick in and what sort of temperatures they will take us to. Nevertheless, there's a fairly high degree of consensus, if not unanimous agreement, that as the temperature goes higher, the probability of these feedbacks kicking in goes up. There is a sufficient degree of uncertainty, but nevertheless, the idea that these feedbacks are there, it would seem to any moderately prudent person, they would say, that it is not worth risking. It's not like, if this house gets burned down, we'll move to the one next door. We are talking about this planet that we live on. We don't have spare ones hanging around that we can move to if this one goes wrong.

We had spoken for almost an hour and it struck me how this type of serious conversation was not part of the wider social and political discourse, not even at the COP. The global leaders outside, absorbed by the fanfare, were not united in collective action because the course of action itself was competing with a narrative of economic growth and consumption. I ended the interview with one last question, asking Kevin what approach he would take if he were given the power to implement a best-chance climate policy.

> I would take the IPCC carbon budgets for 2°C. We're probably on the outside chance of staying below 2°C now, and I'd say that is really the budget range that we have. We then need to make sure that our carbon dioxide emissions remain within that budget range. We then say, how do we split that carbon budget up amongst all the nations that are here? Not the random INDCs – those dodgy voluntary sets of submissions that we are all running around saying how wonderful they are, that have no basis in science and no basis in equity. We would look at the carbon budget that we have for 2°C and then say, how should we divide that up? Now, that would be a very difficult negotiation, but if I was to try to push that forward, I'd want us to be negotiating on the basis of a collaborative view of the wellbeing of our planet and humanity. Not just the wellbeing of people in our particular countries, or particular people in our particular countries. Each country will take their approach to put in the cultural framework that is appropriate for them to stay within the carbon budget that we should be delivering from this event. To me, that would be a reasoned, sensible approach for a caring, sophisticated species. Unfortunately, that doesn't seem to be human, so I don't think we'll go down that route.

Something told me that if the political elites present *had* asked Kevin for his COP21 policy recommendation, the famous gavel, seen thumping down to signify the agreements reached at the closing of the COP, would likely be used to escort him off the premises. It was the first day and Kevin had said he would be there for the full two weeks of the conference, so a loose arrangement was made to meet again. When engaging with experts from any field, we are tasked with searching for sources who offer evidenced and verified information. A particular strength of Kevin and his colleagues' analysis is that it presents evidence of the scientific reality as opposed to political realities. Somewhere, contorted in the midst of both, is the reality that humanity is left to endure.

I walked away, energised by the conversation, towards where I had seen Saleem earlier that morning. He was still there, but this time with another set of cameras and interviewers. He saw me out of the corner of his eye and I signalled I would be back. My stomach was sending me signals it must be lunchtime.

I headed out onto the concourse, where the crowds of people were thickening. The presence of world leaders was making people giddy, and they were crowding around to get a glimpse of their preferred bureaucrats. I walked through the carnage of reporters, each chatting energetically into branded microphones, creating an indecipherable cacophony of multilingual noise. Pushing through the hordes, I eventually discovered a food stall, lured by the sight of rock-hard French baguettes and bitter coffee.

The Global South perspective

What Kevin had said was that 2°C was the best hope. 1.5°C was gone. The difference between 1.5°C and 2°C was real. It was many metres of sea-level rise. It was a far greater supercharging of already supercharged hurricanes and typhoons. Vulnerable nations in the Global South knew that if the agreement that came out of Paris was set to 2°C, then many of their islands and low-lying coastal areas would be submerged. Their people, already among the poorest and most vulnerable on Earth, were going to suffer and, with nowhere to run to, perish.

Not that they were down at heart or giving in. No, they were at COP to be seen, to be heard and to report back to their people what was happening. They were also there to demand justice. They demanded that the developed, high-emitting nations who had caused the crisis must provide them with the funds required to adapt to what was coming and to leapfrog dirty fossil-fuel energy.

Would we show heart and compassion and aim for 1.5°C even if we thought it was impossible? Or would we opt for 2°C, a dangerous and politically palatable number that would likely delay action and drive complacency, whilst committing hundreds of millions, perhaps billions, to suffering?

All these thoughts were swimming around in my head as I ate my sandwich and drank more water and coffee before heading back to wait for Saleem. I arrived fortuitously as he was in that post-interview moment of being unclipped from a lapel mic. Although smiling politely as I approached, I guessed he was frowning inwardly at the thought of another such mic being appended to his lapel. Saleem is shorter than me, with jet-black hair neatly brushed into a parting. He has a kind expression, behind which lies a wealth of knowledge and experience, making him an agile commentator and, when necessary, a harsh critic of UNFCCC progress, or lack thereof.

We began with Saleem discussing the impacts that faced the most vulnerable parts of the world, which this being 2015, were very much underway, even if under-reported.

One of the big problems that we foresee in the long run is that some parts of the world – the dry lands like the Middle East and Africa, very low-lying coastal areas like the Small Island states and parts of Bangladesh – these places are going to become uninhabitable over the next few decades. People will simply not be able to live there any more.

Relocation or migration, as a planned exercise, is something that we have to consider. Where they go is something we will have to figure out. We are already seeing significant relocations from rural to urban. The world population has become primarily urban now, and it's growing in terms of urbanisation.

So that is one axis of movement. Then there are movements across borders. There are already millions of people who go across borders from one country to another in a fairly unplanned and haphazard way. Those are going to be exacerbated by climatic events. We need to think about how we plan those migrations and enable people to go in a systematic way. It is much better to plan and do it systematically than to leave it to chance, because it will happen in a very unplanned and unsystematic way and it will lead to chaos, trouble and conflict.

The images that these words conjure are already being realised, albeit in smaller numbers, yet many of us in the West are proving incapable of showing compassion and empathy. Migrations have been occurring frequently as long as humans have walked the Earth, yet today people in need are demonised and politicised, while narratives around nationalism pour scorn on the suffering of others.

The recurring thought that echoes in my mind is that we in wealthy nations are acting as though we are above the suffering. By continuing to live profligate, high-emission lifestyles, we are saying it is okay if it only impacts on those people far away. Many of us who lead comfortable lives look the other way. Perhaps we recycle and consider that our environmental debt is settled. This indifference extends from us into our social groups, which, in turn, define our values and expectations from our community and society. Politicians respond to these signals, and the outcome is the difference between a policy that works for us and a policy that condemns us. Confronting this systemic injustice has been Saleem's challenge over many decades. Here at COP21, he has pushed to keep 1.5°C in the text of the Paris Agreement, despite being aware of the diminishing chances of achieving it.

> Vulnerable developing countries, like the Small Island states, are arguing for the target to be set at 1.5°C, even if it is extremely difficult, because we believe that it may be difficult, but it is not impossible. We are currently heading for well over 3°C – that is called the business–as–usual scenario. We now have promises from over 150 countries to reduce emissions. If you take those promises at face value and you add them all up, it will take you to 2.7°C. So, that is a long way to go to 2°C and even further if you want to change the goal to 1.5°C. The pushback we get is, 'Be realistic, it is not going to happen.' Well, then we have to live with the consequences that hundreds of millions of people living in the most vulnerable areas are being written off. So, what the leaders in Paris are saying, if they accept a 2°C global goal, for the poorest of the poor and the most vulnerable is, 'We are not going to be able to protect you; we are going to protect ourselves, but we are not going to protect you because it is too difficult for us to remove our emissions,' and that is a very bad message for world leaders to have here.

I thanked Saleem and packed up my stuff. I couldn't help feeling a bit sullen. These were big numbers: billions of tonnes of emissions, hundreds of millions of lives. Other numbers were small, but with huge implications. Mean global warming of 1°C was causing chaos. A mean global rise of 1.5°C was a crazy high amount, and 2°C+ would likely set off irreversible, self-reinforcing extreme climate feedbacks. Our current trajectory was double over crazy-high. All because of greed, misinformation, and an insatiable thirst to consume all we can.

As I glanced around, I was struck by the huge volume of young film crews and general media. Many seemed to film random walkabouts, looking at exhibitor booths as if they were alien specimens, worthy of capturing for the history books. There was much hype in the build-up to Paris, with statements that this was to be the most important summit in human history, deciding our collective fate. These recorders of each passing, granular moment must have felt the weight of this assertion, as they recorded it all for the benefit of grateful future viewers. I had my doubts about it, and yet, as I walked around, the mood was upbeat, joyous, musical, hopeful, *expectant*. Even those protesting in permitted spaces seemed quite resigned to being saved by the COP's multi-state negotiating process.

I headed back to my workstation in the media centre and began downloading the footage to my hard disk. Scott walked over and said that the Canadian climate communicator Paul Beckwith was arriving in Paris and would be at the COP in the morning. Scott had organised a badge for him and he was going to participate in his Climate Show series. Beckwith was particularly focused on the melting Arctic and was keen to discuss geoengineering techniques to try to protect the reflectivity of this rapidly diminishing ice cap.

Geoengineering (also known as 'climate interventions') is one of the most divisive topics in the whole of climate discourse. It is a term so overused and often misused that it is mostly associated with dystopian science fiction. Virtually all the connotations are bad, evoking images of Dr Evil-type technologies that could compound our problems and further degrade the biosphere. Voices on both sides of the geoengineering topic hold entrenched views, although many on both sides advocate further research, if only to prove that it *doesn't* work.

I worked for another couple of hours, transcribing the two interviews, before Mike and Lizzie interrupted to suggest a drink. The crowds inside the COP had thinned out as most people headed off into the night to indulge in countless events arranged throughout Paris as satellites of the COP. Despite the subdued ambience of Le Bourget at this hour, there were still many people dotted around, working, chatting, moving from place to place. It was time for us to emerge from the theatre of the Blue Zone and seek sustenance.

We took the shuttle back from Le Bourget and dumped our stuff in our rooms before meeting at the Place2Be bar. Scott had given Mike the name of a brasserie for us to meet in to discuss tomorrow's schedule. Scott's preferred brasserie was a dour affair, located a minute's walk away from the Place2Be. On entering, we saw him sitting in a raised seating area, forking a salad, looking pale and exhausted. The fatigue was infectious and the food certainly looked tired. I decided I would eat from the bar menu back at the Place2Be.

Scott said, 'Paul Beckwith was just here. I got him his COP badge and he'll be inside tomorrow to help with social media. James Hansen is also arriving tomorrow and I want to be ready. Will you all help me to make it a success?' His voice was ebbing away and sounded almost pleading.

Mike chimed in: 'Yes, absolutely. I'm really looking forward to seeing Hansen speak. We'll be ready. Do you know when and where?'

'He gets in around lunchtime and will arrive at Le Bourget a couple of hours afterwards,' Scott said. 'I want you to film me walking him in and then into the press-conference, which will be packed out. I'm organising press releases to send out via the UN media database. If you are all in early, can we organise ourselves?'

'No problem!' we all agreed.

Scott continued: 'Keep tomorrow night free as well. Naomi Klein and Hansen are doing an evening event at the Place2Be in the room downstairs. Get in early 'cos it is going to be packed. It's been organised by the Citizens' Climate Lobby and they are going to let us film the event.'

I didn't know that Klein was going to be in Paris. I was aware that Hansen and Klein disagreed on the role nuclear energy should play as a low-carbon-emitting transition fuel, based on conflicting research. I was more familiar with Hansen's views than Klein's and looked forward to comparing their two perspectives. While walking back, Mike suggested I interview Beckwith in the morning. He is a well-known communicator on a range of climate issues, and if Mike and Lizzie took care of the technical stuff, I could focus on the questions. I agreed and said goodnight before heading to the bar for a beer, sandwich and French fries.

Back to the COP

I awoke at 6.00 a.m., put on my cotton trousers, shirt and suit jacket, and ran down for breakfast. The usual COP crowd was assembled. I drained two cups of coffee, knowing this would be the staple of my day, and, with bag on back, made my way swiftly into the Blue Zone. Not long after I'd sat down, Scott appeared at my shoulder, quiet and serious, his face tensed into a pale state of permanent concentration. Beckwith arrived and I said hello. We had exchanged a few emails over the preceding years, but never actually met. With short, neatly cropped but scraggly hair, glasses and a day or two's worth of stubble, he presented himself with a reserved, cheerful demeanour. He was trying to discuss his participation in the Climate Show, but Scott was only half listening, preferring to focus only on the Hansen presser later that day.

The list of participants throughout the COP was growing. The Climate Show may have had an air of surreality, but it was a big draw for participants and also an effective method for pooling expertise and getting them to respond in real time to events. It was Scott's hustler instinct that inspired him to mould the resources of the UNFCCC into a personal suite of influence within the shaky walls and bouncing conference floors of the COP's Blue Zone enclosure.

There exists a primitive nature at the COP, where money may be an important factor in the main negotiations, but bartering on the floor is usually for different forms of currency. Many of the thousands of people marauding around wear badges that do not have any connection to them outside the restricted area. This gaming of the system is a very human element and one that creates space for people of all stripes to create their own platforms. Scott's mastery of the systems to his own benefit gave him highly prized curatorial powers in the press-conference area. Over the successive post-Paris COPs, his appearances and performances would peak and then vanish, but other emerging groups would also learn to commandeer the available resources to achieve similar ends.

Scott was now in a state of high tension, his face stuck in an impatient scowl, issuing orders to just about everyone. Hansen was arriving; Scott wanted total dedication from us all. I was partially off the hook because Mike and Lizzie were there to get the main footage. I said I would shoot cutaways of the arrival scenes but Mike had intimated they would have plenty. We had a couple of hours to kill and I was working on the footage from the previous day and monitoring the influx of emails from the UN press release list streaming into my inbox. The UK Green Party politician Caroline Lucas MP had agreed to give an interview on the Friday. That was great news. Caroline was the UK's only Green MP, demonstrating steadfast determination and stamina to keep going, pushing for greener policies and an environment-friendly agenda.

While I was working, Beckwith appeared at my shoulder. Looking around, I noticed that Scott had disappeared. The whole area seemed much calmer in his absence, so I got up for a chat. 'What have you got planned?' I asked.

'I've agreed to do a Climate Show with Scott,' he said. 'I just want to discuss the dramatic changes in the Arctic region and that we need to cool the Arctic.' He said that he was concerned that the Intergovernmental Panel on Climate Change (IPCC), the body of scientists that informs the UNFCCC, was underplaying the risk in their recent Assessment Report 5 (AR5) that had been published in 2014. This was the scientific research that was created to inform climate policy – if politicians would heed the warnings. I asked him to elaborate on the context for his concern. Beckwith explained:

> The Arctic is warming extremely fast and that lowers the temperature difference to the equator, which makes the jet stream move more slowly, become stuck and increase the frequency, severity and durations of extreme weather events. When we lose the sea ice, the Arctic is a much darker place. When it is a much darker place, there is much more absorption of solar-radiation; therefore, there is much more heating in the Arctic. This heating, both in the atmosphere and in the ocean, is a very serious thing. In the ocean, the heating will thaw out the clathrates on the ocean floor. So we're looking at a potential for very large methane release.

The European Commission energy website states that methane is eighty-four times more potent as a greenhouse-gas than carbon dioxide over a twenty-year period. I asked how the impacts he referred to fitted within the 2°C warming boundary that was being pushed for in Paris. Beckwith replied, 'Two degrees is a global average temperature. The actual temperature rise in the Arctic is much, much higher. We've had about one degree Celsius of warming over the pre-industrial period as a global mean average, but the Arctic temperature rise has been about four or five degrees over that same time period.'

When I asked him what he thought the outlook for the next fifty years looked like, he said, 'The first year, the ice would take longer to reform and wouldn't be as thick in the winter. So within a few years, it is very probable that we would have an ice-free Arctic for several months of the year. Then, within a decade, all year round, and then we skyrocket up to a much warmer temperature.'

In 2023, the temperature has risen about 1.2°C above the pre-industrial period and the climate is breaking records and causing destruction day by day. To talk about going over 1.5°C sounds completely insane. Yet, here we are, trying to comprehend humanity existing in greater than 2°C scenarios.

'The biggest threat from rapid warming, or abrupt climate-change, is the threat to food,' Beckwith continued. 'We have to grow food, and what happens is we get areas with torrential rains that destroy food crops.'

In 2023, researchers, led by Dr Kai Kornhuber from Columbia Climate School, published their findings in *Nature Communications* journal, focusing on the climate risk to global food security. Kornhuber summarised his findings on the climate-focused website *Carbonbrief.org*, writing, 'When extreme weather events happen simultaneously – such as a heatwave and drought at the same time – this can negatively affect crop yields in major crop-producing regions, known as "breadbaskets". This co-occurrence constitutes a growing food security risk. Such events can disrupt supply chains, decreasing food supply and leading to price spikes.'

This interconnectedness between many parts of the climate system that are physically so far apart is beginning to be better understood. We now know that the frozen Arctic acts like a cooling system for the northern hemisphere and is connected to the stability of the jet stream. When the jet stream becomes unstable, we get persistent extreme weather events that can cause the crop failures that Dr Kornhuber is talking about above.

Beckwith's concern was that these changes in the Arctic represent a shift to a new climate regime that we will be unable to adapt to. In 2015, this was not so visible to the public, but by 2023, extreme weather impacts around the world really stepped up a gear. Beckwith outlined his three-legged-stool strategy to address these challenges. The first leg of the stool was obvious: reduce our carbon emissions to as close to zero as possible. That was the COP's job. The second leg was carbon-dioxide removal. Although the IPCC and UNFCCC like to use 1.5°C and 2°C global average temperatures as their goals for limiting warming, many scientists prefer to use atmospheric concentrations of greenhouse gases when discussing the issue. Beckwith stated that even if we lower emissions, we would still see a continued temperature rise due to the high level of greenhouse gases in the atmosphere.

This point was reiterated by Dr James Hansen at another press-conference outside the Blue Zone. 'I have a paper that argues very strongly that 2°C is a disaster scenario,' Dr Hansen pointed out. 'For one thing, if we look at the Earth's history, the Eemian, the last interglacial period, 115,000–130,000 years ago, was at most 2°C warmer than the pre-industrial period, probably somewhat less than that, and sea level was six to nine metres higher. So that is a crazy target to have. I prefer not to use temperature as a criteria because we can now measure the Earth's energy imbalance. What that tells us is that we would have to reduce CO_2 from about 400ppm [parts per million] to

350ppm. We really have to restore the planet's energy balance if we want climate to be stabilised.'

Reducing atmospheric concentrations from today's 422ppm (in 2023) means removing hundreds of billions of tonnes of carbon from the atmosphere and storing it somewhere. Beckwith continued, 'You identify the sinks in nature that naturally remove carbon dioxide from the atmosphere and you try to enhance those processes. So you use nature to help you. For example, by reducing the cutting down of trees and replacing them with crop lands. Trees store a lot of carbon. Also, the oceans are a vast store for carbon.'

Large-scale carbon removal proposals remain in the very early research stages of development and nothing yet exists that can be scaled up to a significant degree. The three-legged stool was not even in the prototype stage.

I replied, 'That was the second stool leg?'

'Yeah, you need three. We need to cool the Arctic. We need to look at "solar-radiation management" techniques, SRM techniques, to do that on a short-term scale, to buy us time to get to zero emissions.'

Beckwith outlined two methods of solar geoengineering for cooling the planet. The first was using remote, unmanned ships that pumped seawater into the air to brighten clouds in the Arctic summertime. These clouds, with increased reflectivity, then reflect a larger proportion of the sun's radiative energy back into space. The shaded Arctic Ocean area would then, in theory, stay frozen after the long, dark winter.

The second method involved pumping reflective aerosols into the stratosphere, spreading across the whole globe, which would create a far wider cooling while the light-scattering aerosols remained in place. This aims to mimic large volcanic eruptions that pump enormous amounts of reflective particulates into the stratosphere and have been observed to significantly cool the Earth. I raised a concern that I had heard from scientists who are very concerned about these proposals. Being a dynamic system, whatever forcing is made in one area can create unforeseen changes (teleconnections) in other locations. One such consequence could be to turn off the Indian monsoon. Another might be to cause serious damage to the ozone layer, also located in the stratosphere.

Beckwith anticipated my concern. 'Whenever the topic is discussed, it's always what will happen if we do it? Just as important, in fact more important, is what will happen if we do not do it. If we do not do it, extreme weather will ramp up, and what we consider extreme weather events now will become the norm.'

All climate interventions fall under a general banner of geoengineering. This term comes with a lot of negative connotations. One persistently valid concern is called the 'moral hazard' argument, where even the discussion of geoengineering provides fossil-fuel-producing nations and corporations with an excuse to keep polluting, discarding any sense of precaution given that the technologies neither exist and, if they did, may not even work. I asked Beckwith if he really thought this was where the research focus needed to go.

'We cannot discount these methods,' he commented. 'I think they will all be required when people start panicking about the extreme and abrupt climate-change. Then they'll be calling for these methods, so we better know what we're doing when that happens.'

Almost a decade on, this controversial view is finding growing support despite being confined to a relatively small geoclique and will continue to be raised as extreme weather increasingly impacts on our lives.

2

COP21, Paris, 2015, Part 2

'Half-assed and half-baked!'
The clock ticked. Lizzie came over to say that they were going to get ready to record Scott and Hansen. I picked up my stuff and followed them out. My phone buzzed. It was a message from Scott: 'HANSEN IN. WE ARE ENTERING THE MAIN WALKWAY. ARE YOU IN POSITION?'

I got outside and was overwhelmed by the number of people moving quickly in every direction. I wasn't sure I would even see Scott and Hansen. I saw Mike and Lizzie waiting by the doors, and Mike suggested I move across with my tripod and film the opposite angle. I was a bit more removed, but that was fine. About a minute later, I spied the familiar Indiana Jones-style hat that Hansen wears as he sauntered in our direction. He is surprisingly tall, meaning his hat rode above the heads of the crowds of people. Walking by his side, Scott came into view, navigating the hordes and steering his guest towards the press-conference suite. This was Scott's top guest, and he was certainly looking upbeat.

They entered the building. There was a little fuss at the door as a few members of the press realised that Hansen was here. He is an open critic of the COP process, so there was great interest in hearing his view on the proceedings. I stopped filming as Mike had everything under control. Also, once inside the press-conference room, the multitude of professional cameras and sound-recording systems could capture everything. The crowd was now pretty big, as Hansen was causing a stir. This was Scott's ace card, and he appeared, for a rare moment, to be feeling positive.

People filed in to the press-conference room. Scott and Hansen took to the stage. The show opened with the ten to fifteen seconds of surreal music and Scott thanked the Buddhist foundation for hosting the press-conference. He then laid down a few house rules, thanked Dr Hansen for his presence, thanked the audience, and introduced his guest. The build-up was complete and Hansen signalled for his first slide to be displayed.

For a long time, he has been prescribing a pathway out of fossil-fuel dependency that relies on a carbon tax that is collected from the sale of coal, oil and gas. The tax collected would then be redistributed back to people in society as a form of dividend. The carbon tax (or fee) collected would benefit those who used less carbon and, with the removal of fossil-fuel subsidies, would allow for low-carbon alternatives to be more competitive, while compensating people for the cost of transition.

The problem was that no government was paying any attention. To work, it would require at least two very large blocs, like the US and China, or the EU and China, or all three, to adopt it. Any country not imposing the same rules in their own state would be subject to carbon taxes at the border. The full proposal had the support of economists as both a stimulant to the economy in terms of an equitable pathway out of fossil-fuel consumption and a dividend to help ordinary people with the transition.

Needless to say, the power of the fossil-fuel companies meant it stood next to no chance of being implemented. What we got instead was a conference where nations are invited to *promise*, with fingers crossed behind their backs, to bring down their emissions voluntarily over a time frame that they thought would work based on short-term political concerns or objectives.

Hansen, with an exasperated expression, quoted Christiana Figueres, who had said: 'Many have said we need a carbon price. An investment would be so much easier with a carbon price, but life is much more complex than that.' Dr Hansen then added, 'So, what we are talking about instead is the same old thing that was tried in Kyoto, asking each country to promise, "Oh, I'll try to reduce my emissions. I will cap my emissions or reduce them twenty per cent," or whatever they decide it is they can do. In science, when you do an experiment and you get a well-documented result, you know if you do the experiment again, you are going to get the same result. So, why are we doing the same thing again?'

Describing the Paris Agreement, he said, 'You know, I don't like to use crude language, but I learned this from my mother so I'll use it anyway. This is half-assed and half-baked! There is no way to make it global. You have to beg each nation.'

Scott's man had taken a dagger to the helium-filled COP balloon. The conference centre at Le Bourget was full of jubilant, expectant people from all over the world – excited people from some of the poorest nations, people who knew that climate-change was life-threatening. They came here to be saved by politicians. Hansen made it clear this was a fraud. The deal had no teeth and, ironically, with the much-lauded and loved President Obama in control, the president who led the inaugural chant of, 'Yes, we can!' we were sailing into the burning winds of climate catastrophe with our sails in tatters and our crew high on hallucinogenics.

Hansen's position was clear: the Paris Agreement needed the carbon tax to be deemed a success. The tax was his definition of teeth in an otherwise toothless deal. Hansen ended his main presentation saying, 'I hear all the ministers and the heads of state are coming here planning to slap each other on the back and say, "Oh, we are really doing great and we are going to solve the climate problem." If that is what really happens, then we are screwing the next generation and the following ones because we are being stupid!'

He ran out of time for questions because the thirty-minute timing of the press-conference is so strict, but as he emerged, a pool of press encircled him, and he answered questions for over an hour, literally pinned to the wall as if in a schoolyard confrontation. It was in this pack that he gave one answer that I think explained much of the complexity of the climate problem and how it is ingrained in our values. One journalist asked if he thought all young people should be told not to drive cars or fly. He replied simply, 'You are not going to get people to change by telling them to stop aspiring.' Values and aspirations have been off-limits when we consider our personal contribution to this wicked problem, but, as we head into the mid 2020s, that is changing.

Outside, Scott was feeling good. I was pleased, too. There was something cathartic about hearing what Hansen had to say. This was a scientist who had issued the early warning to the US Congress in 1988, and here he was, twenty-seven years later, calling out the pattern of failure that dominates these events. His warning this time was to the public: be cautious about trusting what is being cooked up here in Paris.

We all gathered and the mood was high, despite the message delivered not being very upbeat. We packed up and made our way back to Place2Be for the second Hansen show, where he would be speaking at the event with author, journalist and academic Naomi Klein.

Back at the hostel, unencumbered by bags, I made my way downstairs to the bar. I saw Beckwith with a beer and joined him. Feeling the pressure of the day dissipate, I ordered my own. The crowds were building and Scott signalled that we had to come in now if we wanted to watch. The talk was also being streamed into the bar area, but the connection was less reliable and the bar was very noisy.

In the room, I could not get a seat. Mike was in the centre of the audience on a stepladder, with Lizzie by his side. He had the best view in the room at the cost of comfort. Beckwith somehow inserted himself in the centre of the front row and then filmed everything on his mobile at point-blank range.

Hansen gave a similar climate presentation, but also talked more about the energy dilemma. Yes, we needed a carbon-free energy source that would allow us, including the poorest people on Earth, to transition to clean energy, but, he said, solar and wind power would not do that for us. He, along with some other scientists, had announced a press-conference scheduled at the COP the next day to declare that no transition would be possible without including nuclear energy.

This announcement was divisive for many environmentalists. People were not happy. Nuclear was largely carbon-free when it was up and running, but the embedded carbon in the building of plants was significant. Also to be considered is the issue of storage of radioactive materials, which people remain very concerned about. Naomi Klein is one of those who are strongly opposed to nuclear power.

After Hansen had finished his question-and-answer session, Klein emerged from the side of the small stage. She began by lauding Hansen for all he had done for the climate movement, saying: 'I think we all owe such a debt of gratitude to James Hansen as the scientific godfather of our movement, who raised the alarm loudly and clearly so long ago, and continues to.'

After this, Klein shifted to make a nuanced case that was probably among the most impressive I have heard at a COP and resonates much more now with what we are seeing many years later, as policy failures continue to play out. Klein started by making clear that she disagreed with Hansen on nuclear.

> I don't think climate scientists are the best people to tell us how to get off fossil fuels. They're telling us we have to do it. But how is a question for all of us. So, you know, all of our work has to be guided by what we're hearing from leading climate scientists, and when we're debating whether climate-change is happening, I totally agree. This should be a debate between scientists, and it should reflect the reality of the huge scientific consensus. We are in this moment now, where the scientists are saying we must act. We are also in this wonderful moment where leading engineers are telling us we can do it. You know, we have engineers like Mark Jacobson at Stanford University who say that renewable energy is ready to get us to one hundred per cent by mid-century. But the how we do it is not just a job for scientists, that is a job for everybody. That has to include as many people as possible. That has to be the broadest possible spectrum.

The Jacobson reference was key, as he had published research that indicated a total switch to renewables without nuclear, using wind, water and solar, could be achieved by 2050. This was brought up at the nuclear press-conference with Hansen and the three other scientists. Addressing it specifically, Professor Ken Caldeira from the Carnegie Institute summed up their position, saying, 'If you make unrealistic assumptions, you end up with unrealistic conclusions. I think that is what Mark Jacobson has done. The goal is not to make a renewable energy system. The goal is to make the most environmentally advantageous energy system we can while providing us with affordable power. I think a clear analysis of this will be part of this.'

I checked this quote with Ken in 2023 and he said it stands. Around the same time, I spoke with Dr Julia Steinberger, a professor of ecological economics at the University of Lausanne in Switzerland. Steinberger agrees with Klein that renewables can get us to a clean-energy paradigm. She says:

> So, renewable energy is now worldwide, so cheap per kilowatt-hour to make renewable energy. We get good surprises with renewable energy. Solar panels last so long. They're making them as parts of roofs. If you told me that ten to fifteen years ago, I would have said, 'You're smoking!' But they know that these things are really solid, and their lifetime performance degrades much slower than people thought. The efficiencies are really high compared to what anybody thought was possible, and they are easy to make right now. So they're cheap. That's one thing. And wind power was always cheap and it's getting cheaper by the minute. Also, we need less energy to do stuff than we used to.

Back to Klein's presence on stage. She was asked about the role of politics in the transition to clean energy, and replied, citing the recent change in Canadian politics. Klein warned against optimism regarding the departure of the Harper administration.

> We have a new government but I would say that this is just the beginning. Electing a government that is not actively obstructionist is only the first stage. So, about four years ago, there was the largest wave of climate civil disobedience at the time, where over 1,200 of us were arrested outside the White House for demanding that President Obama say no to the Keystone XL Pipeline, which is bringing oil from the Alberta tar sands. I was arrested with another climate scientist, Jason

Box, who is now one of the leading experts on the Greenland ice sheet. This is happening with our scientists deciding that the work is so urgent that they must act on it and model the urgency with their behaviour.

But I would say in terms of the world of politics, if we look at the Obama who is here in Paris versus the Obama who was in Copenhagen in 2009, it's a different Obama. I would say that the difference is not just that this is an Obama who is leaving office, this is an Obama under pressure from a much more militant climate movement that isn't just negotiating in the back rooms but is in the streets. It is engaging in civil disobedience and has clarified that climate is a decisive electoral issue and is reshaping politics. I think we get the politics that we demand, that we put pressure on. So, as a Canadian, I feel like the worst thing we could do is repeat the Obama trajectory where we all think, 'Oh, we got rid of the bad guy, we can just relax and everything will be taken care of.' We need to put pressure on Trudeau at this moment!

Despite their differences, Klein and Hansen both appeared to agree that politicians were presenting a grand illusion in Paris, tone-deaf to the real needs of the people and the environment that sustains us. The scientists were linking climate and ecology and, seeing the disconnect between their research and the deaf political ears their findings fell on, they were taking to the streets in despair, ready to be arrested.

Klein continued on the agreement being cooked up at Le Bourget. 'We depend on our climate scientists to tell us the truth about what we are doing, but I think we also depend on our political analysts to tell us the truth about what our politicians are doing, and I think there is a kind of idea that people can't handle it, like we have to just give people positive messages. If we tell them just how bad this deal is, they'll get so demoralised.'

As a reminder that protests in Paris had been banned because of the Bataclan terrorist attacks, Klein talked about the broader impact that this was having on the climate movement of concerned people, and the ability of these concerned people to further influence the momentum of negotiators inside the restricted Blue Zone.

I think it does matter that protests have been banned. I think our politicians make the best decisions when they are under pressure. That said, we know that this agreement will not bring us even to where they promised last time.

We are hearing all of this talk of ambition, ambition; and yet, in 2009, these same countries came together and defined dangerous warming as anything above 2°C. Now, if you've read James Hansen's latest papers, you know 2°C is too dangerous. At the time, African delegates marched through the hallways of the Bella Centre and said, '2°C is a death sentence for Africa.' The island nations marched and said, '1.5 to survive.'

It is these governments that define dangerous as 2°C, and here they are with emission-reduction targets that bring us to three or more. So how can we call that ambition when it's actually going backwards? How can we call it ambition when Kyoto was legally binding and then we're hearing that this can't be legally binding?

I met with Christiana Figueres a few months ago, and she asked me, 'Well, what would you want?' I said, what I think would be most helpful is if we abandon this language of failure and success because we know that this is not going to be a real success for the planet. It is actually quite antagonising when this becomes a space for political legacy-building. What we need is serious engagement. What we need is a deal that has room for improvement.

All of this put more flesh on the bones of what Hansen, Kevin Anderson and Saleemul Huq had been saying earlier about the rhetoric of the COP and the reality of what the science is telling us. It also points to the illusion that not enough people in Paris were taking seriously, but which those in power want us to believe, that the politicians have got this covered. Klein continued: 'When Christiana Figueres said, "Never has so much been in the hands of so few," I understand why she says that, but I don't think it's true. They are not the only ones with power, and I think when we abdicate the power and say, "You are the one. Save us," that's when we get ourselves into trouble!'

This really summed up Paris for me, both then and afterwards. We moved through the conference, waiting for announcements and press releases and statements from Figueres. The process resembled nothing like an emergency. It was, instead, a foregone conclusion that *they* were going to do it. *They were going to save us!* The talk finished, there was applause and many of the audience flocked to the speakers to ambush them with more questions. My beer glass was empty. I bumped into Scott, who in turn introduced me to the organiser from Citizens' Climate Lobby. We were just more faces in a noisy, crowded room and I was thinking more about my empty glass. I slipped away into the crowd towards the bar upstairs, where the music was ear-piercingly loud.

With a beer in hand, I headed for a small, empty-looking mezzanine area, signalling to Beckwith, Lizzie and Mike. Outside, I found a small, isolated spot and sat down with some relief. The music was disorientating and, as I sipped the beer, I felt a sense of fatigue creeping over me. I couldn't see the others in the throng, so I finished my beer and headed for bed.

Another COP21 zeitgeist narrative

The next morning, I woke up at 5.45 a.m. I dressed at speed and, with bag in hand, descended in the lift, knocking back a coffee and filling my pockets with fruit and a croissant to go. Of the various passes at the COP, the media or press badge was the best to have, as it provided the holder with more access to interesting events. The negotiations were uneventful, and I could observe other sessions via any of the hundreds of screens all plugged into the various feeds within the Blue Zone.

This morning I was attending an event to discuss 'stranded assets', a term coined by a group of financial analysts called Carbon Tracker. Their findings were published as compelling arguments to put to investors for why divesting from fossil fuels made good sense. The session featured the CEO of Carbon Tracker, Anthony Hobley, a lawyer by training who had moved his career in the direction of climate-change, coming from a much more corporate perspective.

When I arrived at the location of the session, there was a vast crowd at the entrance, which meant getting in would be a struggle. I really didn't think stranded assets were as sexy as all that. Niche, yes; sexy, no.

It turned out that the session was being introduced by former US Vice President Al Gore. In climate terms, Gore is definitely somewhere at the top of the pyramid structure of climate respectability, an establishment grandee on the world stage. His Climate Reality Leadership Project has trained hundreds of thousands of people to become 'climate reality leaders' around the world.

The presentation Gore gave in his movie *An Inconvenient Truth* is the foundation of his Climate Reality training, as attendees learn how to construct a climate presentation that is relevant to their location and culture. It is an interesting concept that has created a network of impassioned climate advocates. In the rush for a place, I dived, like a porpoise, into a seat. Gore began one of his well-crafted, rousing speeches. He asked three questions: *Should* we change? *Can* we change? *Will* we change?

To the first question – should we change? – he began the backstory of climate devastation that has already arrived and is on course to get much worse.

> The television news every evening is like a nature hike through the Book of Revelation. We are still emitting 110 million tonnes of heat-trapping, man-made global-warming pollution every day. The cumulative amount that still resides in the atmosphere traps an amount of extra heat energy each day equivalent to that which would be released by 400,000 Hiroshima-class atomic bombs exploding every twenty-four hours. It's a big planet on which we live, but that is a lot of energy. Over ninety per cent goes into the oceans and that results in the incredible increase in the water vapour that is evaporated from the oceans into the air. The atmosphere holds more water. As a result, when the hydrological cycle continues to move that water vapour over the land masses where it falls as precipitation, the amount of precipitation is much increased.

This last point is of enormous concern. The increased water vapour is driving the incredible thunderstorms, or atmospheric rivers, that we see around the world, causing flooding and destruction of urban and natural habitats. These are now with us and, although we have always had heavy downpours, they are getting much more violent.

Living soil represents a huge store of carbon, storing roughly twice as much as the atmosphere globally. As the level of energy in the atmosphere increases, it causes more moisture to evaporate from the first few centimetres of soil. This destroys the microbial characteristics of healthy soils, leaving them as lifeless dirt. If we lose the soils, as with the oceans, we lose the climate battle.

The second question of 'can we change?' was answered by an overview of the growth of investment in renewables and the falling price of clean energy, which was increasingly competitive with the cost of fossil fuels. It was not there yet, but, he forecast, when renewables cross a threshold, we should see an energy tipping point or paradigm shift.

His third question – will we change? – was geared around the signal to investors that in a warming world, with global agreements in place, limiting emissions to avoid the catastrophes outlined by question one, the hundreds of millions and billions that are being invested into these destructive industries will be lost. The infrastructure will become worthless and therefore the capital invested will be stranded. This was the Carbon Tracker risk assessment, and I heard it echoed around the COP more than a few times. Anthony Hobley emphasised that Carbon Tracker was based on the work of seven former financial risk analysts and that the signals they were detecting were not at all understood within the financial industry.

With fossil-fuel companies creating their own propaganda, framing their own polluting products as profitable for decades to come, investors were still piling in with billions, if not trillions, of dollars every year. The risk was now two-fold. First, with global carbon emissions threatening the entire globe, as Gore had said, and second, with the economic risk of a gigantic financial carbon bubble that, when it pops, would decimate the global economy and create an unimaginable collapse of investment funds. Hobley compared the risk to investors as being like that of the cartoon character Road Runner, who only realises he has run out of road when he is several metres beyond the cliff edge.

This analysis was intriguing. Carbon Tracker had created a climate plan in financespeak that could then be presented to the captains of industry, who would, of course, do the right thing. Or would they? The risk analysis, empowered by a Paris Agreement that had the teeth of enforcement, was really essential to bend the steepening carbon emissions curve to within scientific boundaries for beneath 2°C and as close to 1.5°C as possible. Petrostates and the corporations they owned, or subsidised, literally needed to phase out fossil fuels while transitioning to clean-energy production.

This was COP21. So in the last twenty-one years of COPs, no progress had been made in emissions reduction. In fact, if you consider the annual emissions that have been steadily increasing to around 35.7 billion tonnes as they were in 2015, then we have made the problem considerably worse, and the ramifications of this are not showing signs of abating.

An agreement with no teeth was an agreement with no ambition. All the analysis and research demonstrated that greed was now quantifiable in tonnes of carbon, and we, like true nutters, refused to do anything about it.

I left the Carbon Tracker event intrigued by their research, but with a sinking sense of how critical it was for agreements to be legally binding. The tragedy felt baked in.

I rounded a corner into the hall where all the national pavilions were. A large crowd was gathered, with people wearing their indigenous costumes. They were wearing make-up and were full of joy. This was the Peruvian pavilion, and I could see coffee and some food being passed around. Having a camera on a tripod, I was welcomed into the fray and quickly offered a small cake, which I devoured with enthusiasm. Music was playing and people were swaying. It was still mid-morning, so this was a good little calorie boost to stave off any creeping fatigue. As I looked around, considering whether to stay or go, I was offered a small plastic cup of clear liquid, which I assumed was water, so I downed it. It turned out to be Pisco: Peruvian firewater. My face spasmed into a smile and the woman offering it laughed and asked, 'Do you like it?'

'Of course,' I replied. 'It's delicious!'

'Would you like some more?'

She held up the whole tray, and I took two of the cups. I had a feeling that it was going to be a long day, so a few stiff drinks might soften the impact. I thanked her and stood on the edge of the crowd, grinning as people danced and chatted and handed out documents about Peruvian culture. I wasn't sure if it was to do with climate or if it was a touristic travel performance. Either way, I liked it. It gave me a warm glow and a stumbling walk as I sensibly navigated my way to the German pavilion to get a double espresso before heading back to my desk.

Thawing ambition

A little later, I read a tweet that the glaciologist whom Naomi Klein had mentioned in her talk, Professor Jason Box, had arrived at the COP. I sent him a message to see if he was available for an interview.

It was a day later when he replied saying he was in the media centre and could meet. I had watched a documentary that Professor Box featured in called *Chasing Ice*, screened in the Houses of Parliament in the UK. The combination of cinematography and personal story underpinned by hard science made it a hit. It showed the receding of the glaciers on Greenland and in other places, winning the Satellite Award for Best Documentary. Jason had also been arrested protesting outside the White House in Washington and had made high-profile TV appearances in the US, speaking out on the urgent need to address climate-change.

A year before the Paris COP, he had written a tweet saying, 'If even a small fraction of Arctic sea-floor carbon is released into the atmosphere, we're f★★★★d.' It reads as an innocuous statement in the current light, given the accelerating rise of methane from both industry and organic sources like peat bogs and permafrost, but for some reason it was seen as hysterical by some commentators at the time. It also demonstrated the role social media was playing in enabling scientists to express what they were seeing and feeling, instead of being out of sight, confined behind the barriers of research-journal paywalls and technical scientific data.

I recognised Professor Box instantly from the film. He cut a distinct figure with his blond hair and neat, pointed beard. I wasn't expecting him to be so smartly dressed, in a navy-blue suit. He waited patiently while I set up, and we went straight into the interview. I started talking to Jason about his work on the Greenland ice sheet and what he was learning about how it is melting.

The ice sheet is three kilometres thick in places and sits on a submerged bed of rock. The base is being increasingly lubricated by the water flowing down from the melting surface of the ice sheet, through fissures and melt-water tunnels. This is causing the massive body of ice to slide off its bedrock. As it does so, the ice sheet edges forward into a grounding zone where it is no longer sitting on the bedrock but extending over the warmer ocean or ice shelves, breaking up and calving into icebergs.

When the whole ice sheet melts, we can expect up to seven metres of sea-level rise globally. With constantly rising carbon emissions, there are no countervailing forces to stop the melt. So the question is, how fast? As Jason explained:

> The Greenland ice sheet, like any land ice mass, is a threshold system, and Greenland is beyond its threshold. The real point

is, how far beyond that threshold do we push this ice sheet? Do we push it to a 2°C or a 4°C summer warming? There are two scenarios. The business-as-usual scenario will have Greenland warming in summer at the end of the century by about 6°C. Then the carbon-mitigation scenario is about half as much. So it's about 3°C in summer, but it is not losing ice super-fast. It's losing about 300 billion tonnes per annum, which is about just under a millimetre per year. So if that kept up linearly, it doesn't add up to much at all.

A millimetre per year doesn't sound like much, but from the conversations with Kevin Anderson and Beckwith, it felt pertinent to ask if there were risks of tipping points, or what Kevin referred to as non-linear factors. 'Yeah, there must be non-linear effects, and we actually see the shape of the ice loss appears to be non-linear,' Jason said. 'What is really important though, if we simplify it as much as just looking at temperature, as the temperature rises, the rate increases non-linearly. So if we have a climate-change scenario with a plus 3°C summer temperature in Greenland, that is much lower in a reduced-carbon scenario than the business-as-usual scenario. In other words, mitigation matters.'

Speaking with different people throughout the COP, my view of a sufficient outcome was a combination of wanting the COP to be a success and realising that there were too many caveats to enable the process to deliver. As someone whose feet are normally walking across the melting Greenland ice sheet, I wanted to know how Jason was perceiving the Paris vibe. 'Before this meeting, I was thinking a lot of this is just theatre, but when I'm here and I'm networking with people, I'm given more hope because practically everyone here really wants to tackle this problem,' he told me. 'I think nations are taking this more seriously now. Is it too late? It's never too late to care.

Yeah, sure we have lost species, we have lost the chemistry in the ocean. That problem seems to be irreversible, the acidification trend. So, yeah, we are a bit on the ropes. We have still got to stick it out, just for the sake of our kids and nature.'

The 'we' in his last sentence seemed to move away from the politicians and more towards civil society. The limits to the political process had a lot to do with the displaced agency of the populace. If the wider public applied a great deal more pressure on the politicians collectively, then the politicians would move faster and harder. I asked Jason what it would take to get that collective shift.

'I think we are going to see that the more pressure goes on people, the more we will see action exponentiating to meet the also exponentiating climate-change impacts,' he replied. 'That is clear. Our response has to meet that exponential.'

Jason reflected many people's attitude here at the COP, that there was good reason to hope a deal would emerge to reduce carbon emissions and slow the rate of losing ice sheets. Note that even in the best-case scenario, the Greenland ice sheet was still beyond its threshold. It would eventually melt, just at a slower rate, giving humanity and other species more time to adapt to it.

What Jason and I were talking about in terms of the exponential social response to the climate issue was not, as far as I could see, emerging in Paris. It felt suppressed. I am not sure whether it was the shock of the Bataclan terrorist attacks, or, as Naomi Klein said, the abdication of agency. The spirit of the people felt insipid, resigned to accepting a toothless Paris climate deal.

Into Paris with Scott

Back at my desk, I plugged in the camera to download the Box interview. While it was digitising on my screen, I received a text message from Professor Hugh Hunt, who had just arrived from Cambridge in the UK and was now in the Green Zone. The Green Zone is the area of the COP that is a public exhibition space for initiatives, talks, meetings, performances, all on the topics of climate, ecology and innovation to tackle environmental issues. Hugh was in there doing some filming for a documentary on geoengineering. Given that it would take me at least forty-five minutes to get there and I was still working in the Blue Zone, I suggested a beer later. His reply agreeing to this suggestion came back at lightning speed.

Hugh's presence gave me an idea. Earlier in the week, Mike had suggested a climate conversation *if* we could get a couple of good speakers. It occurred to me that Hugh's many TV documentaries recreating feats of engineering coupled with a knowledge of climate-engineering projects would make him a perfect candidate. Professor Kevin Anderson, also with an engineering background, extensive public-speaking experience and fastidious adherence to scientific fact, appearing with him could make an excellent duet. Mike and Lizzie both agreed that it could work. I messaged both Hugh and Kevin, and out of this came a very limited slot of thirty minutes at 5.30 p.m. the following day to record. I said I would prepare some questions that we could use as a guide for the discussion, which I started on right away.

Having finished the structure and transcribed the Jason Box interview, I finally bowed to the effects of hunger and thirst. It was time to leave the COP village and venture into the Parisian night. Emerging from the Metro somewhere near the Bastille, I finally found Hugh in a bar surrounded by fellow Australians, all wearing gigantic blackened angel wings. I decided not to enquire what this was about and instead caught up with Hugh, recounting the events in Paris thus far. His fresh enthusiasm contrasted with my chipped and worn frame of mind.

We sipped numerous French beers. He was curious about what had been happening thus far in the Blue Zone. When I attempted to recall aspects of various interviews and presentations or important announcements, it all seemed to meld into a mentally unapproachable noise. So, I switched course and described the surreality of the place. I found myself recalling the strange mélange of the Blue Zone, where politicos, celebrities, media types, indigenous peoples, business executives and any other percentile of the human race who had skin in the game of planetary life, whether for good or bad, came to parley.

Hugh was looking forward to the meeting with Kevin the following day and was obviously more spritely than myself. Feeling the allure of slumber and having suitably imbibed, we said good night. I departed for the Place2Be.

The next morning, I was transcribing my way through endless footage. Scott appeared at my elbow, asking if I would be interested in getting a cab across town to interview Dr Rajendra Pachauri, the former chairman of the Intergovernmental Panel on Climate Change (IPCC), the scientific body that advises the UNFCCC.

'When?' I asked.

'We have to leave now. Will you help me?' he requested.

'Of course,' I told him, 'if I can ask a few questions of my own.'

'I don't see why not,' he agreed.

'Let's go then,' I said, grabbing my stuff, and we made for the exit.

On the journey across Paris, he asked me what questions I was going to ask. My view was that the COP seemed to exemplify how far the bureaucrats were from the reality of the science. Every scientist I had spoken to said that we were in a really dangerous situation. The tipping-point scenarios were enough of a reason to phase out fossil fuels in wealthy countries with immediate effect. What Dr Saleemul Huq was saying about 1.5°C being the absolute limit for the vulnerable nations should be applied to all of us from a precautionary perspective. I wanted to find out more about what we are committed to already and what it would really take to hold the limit to 1.5°C. Is it even possible?

Second, there was the IPCC itself. It takes six to seven years to produce the reports that were informing policy. Yet, the climate was changing in real time and policy was lagging far behind the IPCC reports. It appeared to be a systemic failure at the very heart of how we are *not* responding to the climate problem. I said I would ask Pachauri about this too, and whether he agreed we needed a more agile approach that matched the pace of the problem.

Scott thought about this, then replied, 'I like the idea of sticking it to the IPCC. However, I need something much more powerful, more hard-hitting, something that *really gets you in your guts!*' he said, clamping his jaw and thumping his stomach. He paused, his eyes on mine. 'Can you write some questions for me?' he asked.

'What?' I said, feeling alarmed, considering his stomach-curdling brief. 'How about the difference between 1.5 and 2°C and his view on where we are now? And can we adapt?'

He already had his laptop out in the cab and was hammering away at his keyboard. I offered a few questions that seemed reasonable, but suggested if he wanted expressive, gut-wrenching answers then he would have to find a way to ask them that would inspire those answers. He nodded.

As we entered the Victoria Palace Hotel, close to the Montparnasse Tower, Scott made for reception to interrogate the staff while I found a quiet enclave with seating where we could film. A message was sent to Pachauri that we had arrived, and we sat there, huddled around Scott's laptop, working on his interview questions. Our interviewee arrived with a spring in his step. He was jolly and enthusiastic. We clipped a mic to his lapel and then I suggested to Scott that he ask his questions first. Scott asked about adaptation to 1.5°C and the likelihood that politicians in Paris could agree to hold to it, to which Pachauri replied:

> Now, some parts of the world are threatened already, and when you have extreme precipitation events, where you get a sudden downfall in a very short period, it is a risk to life and property. Also, it is a serious problem regarding the availability of water. So, what I'm saying is that, even with 1.5°C, the impacts are going to be pretty serious and will require a huge amount of effort, resources and adaptation. So it is a value judgment. Human society has to decide.

Scott then asked about the Arctic sea ice disappearing and whether Pachauri thought that could be a human extinction event. After he posed the question, he leaned back with a smirk. Pachauri was not fazed. He said, 'It is true that the Arctic is heating up much faster and there is uncertainty around what impacts that will have on the global climate system, but there is no evidence, I believe, that we are facing extinction.' And then he smiled, almost teasingly, and said, 'Yet!' I am sure he was feeding his questioner the stuff he knew he wanted. It seemed to hit the spot.

Scott gulped hard, as if he were swallowing a box of pins, and asked him, 'Are we close, Doctor?'

'Look, there are an awful lot of people on the planet, so no, I don't think we are close,' Pachauri answered. 'It is probably better to focus our energies on helping those who are going to suffer most, rather than worrying about extinction.'

Scott was placated. He had the 'yet' answer, and he had the word 'suffer' in the second part. Lots of potential there.

I now asked about the IPCC process, and Pachauri seemed to agree that the current system was far from adequate. 'I think it would be difficult to envisage a climate response that was in real time, but we need to be more responsive. Perhaps moving reporting to monthly, or at least annually!'

This issue around the regularity of the IPCC reporting continues to disconnect the climate-change issue from effective policymaking. Operating this way, humanity would always react to extreme climate impacts, as opposed to adapting to what we know is coming and being resilient.

The interview ended, and we made our way back to Le Bourget. It had taken a whole morning, but I felt it had been worth it. Scott seemed pleased with the effort. His Climate Show was going well too. The session with Hansen had been a success, with his key message about any Paris Agreement needing to include a carbon tax being reported in *The Guardian* by Oliver Milman, which read: 'According to Hansen, the international jamboree is pointless unless greenhouse-gas emissions are taxed across the board. He argues that only this will force down emissions quickly enough to avoid the worst ravages of climate-change.'

Milman, in a separate article, then quotes the response from the then US secretary of state, John Kerry, who said, 'With all due respect to [Hansen], I understand the criticisms of the agreement because it doesn't have a mandatory scheme and it doesn't have a compliance-enforcement mechanism. That's true. But we have 186 countries, for the first time in history, all submitting independent plans that they have laid down, which are real, for reducing emissions. And what it does, in my judgment, more than anything else, there is a uniform standard of transparency. And therefore, we will know what everybody is doing.'

How 'real' these independent plans were for reducing emissions would be seen over time, as would the gap between political rhetoric and the success of the Paris Agreement.

Scientific duets

Mike messaged to say they were heading over early to set up, as Kevin had to be away by 5.30 p.m. for a dinner at the Paris City Hall. I packed up my filming equipment and headed for the Green Zone, which was further away than it seemed. Although the Green Zone was next door to the Blue Zone, I had to walk from the media centre to the exit, through the security perimeter, across to the Green Zone and through another security check – a system as rigorous as any airport. I was walking at speed to get past the heavy foot traffic, which was at its peak at the end of the day.

Near the exit, having walked through a field of tall columns, each branded with a flag of a member state of the United Nations, I found a lot of people crowding around the shuttle buses. I ploughed through and joined the fast-moving queue into the Green Zone. I was getting pretty swift at the security scanner too, stripping down to my unfettered self and then emerging on the other side to reassemble.

The Green Zone is often unaffectionately referred to as the Greenwash Zone because it's full of optimistic visions of a future world, with proposed tech, presentations, performances and a great deal of corporate branding. There were also large academic institutions, government agencies and interactive experiences. The buzz reminded me of a consumer trade fair where the public, children included, can come and rejoice in a narrative of techno-futurism. It was alien territory and, as I lumbered through, straining with my heavy bag and general fatigue, towards the furthest corner of the furthest hall, I realised Mike had found the remotest point of the entire COP complex in which to record the segment.

As I entered the second adjoining hall, in the final leg towards his location, I had a sense that I was being followed. I turned to see Scott striding in my wake. 'Hey, Nick!' he yelled.

'What are you doing?' I asked.

'I followed you,' Scott explained. 'I hear you're doing some filming with Kevin Anderson and Hugh Hunt from Cambridge. Mind if I join?'

He had the expression of a detective hot on a trail.

'If you like, but we don't have much time,' I told him. 'It's all last minute.'

I turned and continued to walk, while he paced along at my heels. When I reached the filming location, Hugh was already there and came over, smiling. He introduced the two filmmakers working on the geoengineering documentary, who were setting up to take some additional footage and pool it.

'Hey Nick, how are you?' Hugh said in his Cambridge-softened Australian accent. 'This is Nadja and Alba, who are working on the film about geoengineering.'

We said our round of introductions and I noted it was a few minutes to 5.00 p.m. Looking back along the way I had just walked, I saw Kevin making his way towards us. Mike and Lizzie were setting up. Throughout the Paris conference, this was the only time that Mike had had a chance to properly choreograph a session. The others had all been much more on the fly. He took my camera and positioned it to provide his cutaway shots. It looked like a baby-cam next to his giant bit of kit. All of this prep would eventually show through in the finished recording, presented with more polish than usual.

Kevin arrived and reminded us politely about the time constraints and said that he hadn't realised how far inside the

Green Zone we actually were. Being so deep inside the Green Zone, with only thirty minutes available, it was now obvious he was going to be late for his next appointment in the centre of Paris. Mike took over and got the guys seated. Alba and Nadja were filming additional cutaways off to the side. When Mike announced we were ready, Scott picked up my laptop with the outline of the discussion on it and said he would like to feed in the questions. We didn't have time to argue about it, and the questions were already written out. Mike said he could just edit Scott out, keeping the focus on the two. Without a moment to lose, the only way was onwards. At Mike's signal, we began the session. Kevin began by saying there was now widespread acknowledgement among the social groups in society who are both influencing the politics and perpetuating the problem that the climate science is right. He continued:

There is still a huge disconnect between what the science is telling us and what that actually means for us in our lives and for the policymakers.

Everyone – the climate scientists, the engineers, the social scientists, the politicians – there is a real cognitive dissonance, there is a duality within us. This is not about different groups of people. It is within the same people. We acknowledge the science on one side, but we don't like the repercussions of what that means. That, to me, is the disconnect.

The science has been very clear. Now we have these carbon budgets, and the problem with carbon budgets, they have really big policy implications. The old framing was a 2050 time frame, when we could all just rely on engineers solving the problem in 2035, 2040 and 2050. The carbon-budget approach brings it down to what do we do today and what do we do tomorrow? So the political repercussions come out of the science, and when we look at that, the big problem is time.

So we can come up with wonderful solutions – we can electrify cars, we electrify heating and we can electrify some of industry – but that will take two, three decades to do that. Even if we have a Marshall-style plan, which is what I would suggest we need, to try to make those sorts of transitions. But that will mean we are still breaching our carbon-dioxide budgets. That is why we have come down to the very difficult political part of it, which is, we need to reduce our energy demand. It's just an outcome of the maths.

One reason civil society often has to fight for environmental justice is that politicians often create policies with economic growth in mind, and then look for ways to bend the evidenced studies to fit the objective. Both Kevin and Hugh use the numbers that come out of the science in order to demonstrate how out of balance policymaking is. In a saner world, numbers would be used much more honestly, to ensure that we faced uncomfortable truths earlier on. In the real world of climate nutters, the numbers are out of control. Hugh gives us a taste of the numbers we are dealing with:

We are, as a planet, burning fossil fuels and producing about 35 billion tonnes of carbon dioxide every year, with a population of 7 billion; divide that 35 by 7, and that is 5 tonnes, on average, of carbon dioxide per person. Now, if I ask every individual on this planet to handle five tonnes of waste, whether it be sewage or whether it be garbage from the kitchen, five tonnes per year, that is massive. We do not produce five tonnes of household waste per year. Yet somehow we are allowed to produce five tonnes of CO_2 waste per year.

Now, if we think about how we manage our lifestyle to reduce our CO_2 waste to say one tenth of that, well, that means you have to do essentially one tenth of what you currently do because everything we do is written in this CO_2 ink. Isn't this the way it is?

Hugh's crude calculation, averaging out emissions between the global population, is one way to visualise the vast amount of carbon pollution that is emitted. Kevin's response to this was to cite researchers Piketty and Chancel, who looked deeper into the 'who consumes how much' question, and reveal what this data shows us.

Averages really can be quite misleading here. If we look at the top one per cent in the United States, it says that they're emitting over 300 tonnes per person, and I think the bottom one per cent of Nigerians are about 0.15 tonnes.

If we look at the top ten per cent of people, who are responsible for fifty per cent of global emissions, their emissions are twenty-five to thirty tonnes per person. So what we're talking about is a relatively small percentage of the population who will have to make rapid and radical reductions to their energy consumption, and hence their emissions, in the short to medium term.

The problem is, they are the policymakers. If the rest of the population is going to have lives that are worth leading, then that ten per cent have to make those sorts of changes. That is why we come up with any spurious technique we possibly can to avoid the requirement of people – like us professors and those of us in terms of income levels, which relate to our emissions – having to make these sorts of radical changes to how we live our lives over the next two decades. Even if it is for a short time whilst we put this low-carbon energy supply in place.

Sticking with the numbers, but moving from attribution of emissions to the subject of climate repair, Hugh takes up one of the legs of Beckwith's 'three-legged stool' argument: removing carbon from the atmosphere. Instead of seeing this as science fiction, Hugh uses an analogy dating back to pre-Islamic mythology to frame the challenge:

> One genie we have to pull out of the bottle is to get CO_2 out of the atmosphere. We have put, in the last 250 years, getting on to a trillion tonnes of CO_2 into the atmosphere. What does this number mean? It is so big, like how many grains of sand are on a beach. It is just uncountable. But it is a big number. We've got to get the CO_2 out again. Over the next century or two, it is naturally going to make its way into the oceans, and the oceans are becoming acidic. From what we know about acidic oceans, shellfish can't make their shells and the delicate ecology of the oceans is in an extremely precarious position. So that is one genie we have got to pull out of the bag.

Kevin responds:

> There are two genies here. Both are hugely challenging, but the first one is the social challenge of us actually reducing our energy consumption whilst we put in place this low-carbon energy supply.
>
> The other genie is this highly risky approach of assuming that some technology in the future will suck the CO_2 out of the air, and also in the interim, we keep our fingers crossed that these feedbacks that you were talking about before – this idea that the permafrost melting and releasing methane into the atmosphere – that these feedbacks actually won't start putting lots of other greenhouse gases into the atmosphere and then what we do is almost irrelevant. So there are two genies there.
>
> One, we have some chance of actually realising it, although it will make things very challenging for us. The other one is

this Dr Strangelove technology in 2050 or beyond, and our fingers crossed in the interim that we will not get these sets of feedbacks.

Hugh then raised the historical comparison of the hole in the ozone layer created by chlorofluorocarbons (CFCs). Humanity was able to respond effectively and ban CFCs used in refrigerators. Hugh's point is that humanity has become complacent and overconfident that science can fix climate-change with a neat engineering solution. If so, where is it?

Kevin expanded on his point:

This is a huge issue that we think we have these historical precedents, like the ozone issue, like acid rain. We think now we can do the same thing with carbon. The problem with carbon, it is in the dyes in my shirt. It is in the ship that brought my shirt here. It is how we travelled to this event. It keeps the lights on, it is keeping your computer running. Carbon is completely pervasive. It is in every single facet of our daily lives. We have no historical precedent of ever being able to change society, in relation to removing carbon out of it, within two to three decades.

As we steered towards the end of the conversation, Hugh looked up, gesturing with his arm at the booths and pavilions, asking, 'Why is everyone so optimistic?'

Kevin responded:

Yeah, I mean, we are certainly not prepared for the sorts of temperatures we are heading towards, and the impacts, at all. But again, part of the optimism comes from rich people in the northern hemisphere who think that we can buy our way out of this. I think it is quite a sad reflection on us as a species that we have not been able to look at this issue, which is probably one of the first fully globalised issues we have had to address, and we are failing at it, partly because of our lack of humanity.

Amid all the noise of COP21, this recorded session probed the complexity of solving the climate issue, with enough humility to admit that what we are striving to grasp is now out of reach. This proved to be an informative session, very well produced by Mike and Lizzie and very well presented by Kevin and Hugh. Scott also read out the questions as intended, so I was relieved not to have to make a fuss with so little time to play with. An evening of food and refreshment beckoned.

Short-term politics

Two interviews that stood out to me from Paris were with two very different elected officials: one from Alberta, Canada, Minister for the Environment and Parks, Shannon Phillips, and the other from the Green Party in the UK, Caroline Lucas MP.

I met with Shannon in the area where I had filmed with Professor Box. She arrived wearing an orange dress with a black blazer and neck scarf, her shoulder-length, dark-brown hair swept back behind her ears. Beneath her immediate impression of confidence, I detected that she had a slight wariness of the media. In my questioning, I stuck to the guides of what I had learned from climate scientists, placing them in the context of the future lifespan of the oil sands. The science has consistently and unequivocally stated that they must be phased out as quickly as possible. Shannon didn't agree with this assessment, saying:

Well, you know, Alberta is going to be an energy producer for the foreseeable future, and the world will require fossil fuels for the foreseeable future. So, what our responsibility is here is to put in place the policy architecture so that we can lead the rest of the provinces. We can lead in North America, we can lead in Europe. What we have put in place, if everyone did what we are doing, we would get there. You know, the fact of the matter is, we have taken on tough questions. We've exercised leadership in the first six months of our mandate and, you know, we're very proud of that. We live in the world we live in. We haven't built the new world yet. So, you know, we take that pragmatic approach because, you know, there are so many Alberta Canadian jobs that rely on this industry. And we need to be thoughtful and careful and intelligently determined on how we make those transitions.

Although I could certainly sympathise with the jobs issue, including the need to re-skill and develop new clean-energy industries, Shannon did not even seem to acknowledge that the tar sands (politicians tend to use the term 'oil sands' while activists prefer 'tar sands', given the density of the oil) would have to be scaled back. The 'new world' she referred to appeared to be a distant place, far beyond the horizon of near-term electoral cycles.

In 2023, I interviewed ecological economist Professor Julia Steinberger, who has consistently spoken out about needing to get fossil fuels out of the system, not using the soft language of 'a transition' but in the more robust framing of 'a transformation' of our societies. Discussing the fight with fossil-fuel companies, she said, 'I think that a lot of people, from policymakers to media pundits, journalists to climate scientists themselves, really just hoped that the fossil-fuel industry could be convinced that climate-change is real and that they should be on the right side of history and reorient their industry towards renewables and other kinds of products.'

Citing the longstanding betrayal of the fossil-fuel industry, which knew with a great deal of certainty about the climate threat many decades ago, she continued: 'This industry should never have been taken seriously as a partner. They are not energy companies. The solution you get from models where you say, "Oh, well, you make one kind of energy; well, you could make another kind of energy." But that's not how these companies operate; they really identify with their products, and they really identify with petroleum, gas and coal. For them, that is the name of the game. They don't want to switch to the other [clean sources] and they have the political power to enforce dependency on their products – through regulatory means, through state capture or through getting subsidies.'

Climate scientists had stressed the imperative of keeping 80 to 90% of fossil fuels in the ground to hold the climate to a 2°C heating boundary. I asked Shannon, if she accepts that climate-change is an existential issue for millions of people around the world, how could she rationalise the economic benefits of oil extraction against further climatic degradation.

She told me: 'I think in the short term, Alberta is an energy-producing province, so we need to find ways to turn the transition away from high-carbon energy production to lower carbon, constrained and eventually zero, and in many cases, zero.'

'Would that be this century?' I asked her.

'Well, again, we're in pretty early days here, right? We have, as a government, been fairly loath to set lofty targets that we have no policy framework in place to achieve at all. It's all very convenient to talk about what might happen in this century. Our job as a provincial government is to put policies in place so that we know what's going to happen between now and 2020. That's our mandate. That's a responsibility to the people of Alberta.'

As we continued to speak, Shannon oscillated between enthusiasm for policies that sounded to me wholly inadequate to fit within the policy goals being discussed in Paris and being uncomfortable when attempting to articulate a world beyond the era of fossil fuels or, specifically, beyond the time frame of the next election cycle. This was underlined when I asked if she could ever see a time when they would try to close the oil sands down. She said, 'You know, I don't think that is fair. I don't think it is a fair question.'

'Even as an aspiration?' I persevered.

'I don't think that anyone wants to shut down their main industry as an aspirational goal of government.'

'But if it is doing so much damage and the rest of the world is trying—' I went on.

'I mean, we are beginning the process of lowering emissions out of the oil sands as well. As I said, before, we are an energy-producing jurisdiction, and that is the underlying piece of the economy, right?'

This interview demonstrated how policymaking can be completely detached from the scientific reality that is warning us we must change course. Hugh Hunt and Kevin Anderson had previously discussed the complexity of emissions reduction over several decades. Shannon did not want to talk about beyond 2020. The issue was simplified down to the near-term electoral cycle and talking about actions that would have zero impact in reducing Canada's incredibly destructive fossil-fuel industry.

I should not have been surprised at this response to the questioning, but experiencing it first hand, including the uncomfortable body language, demonstrated for me the entrenched nature of fossil fuels in our societies. To me, it felt like the dark shadow the industry casts over our society. After the interview, I packed my stuff away in a locker to store it overnight. Beckwith came over and asked me what I was up to. I suggested a beer, to which he readily agreed.

'A job of work'
I rose early the following morning to get into the centre of Paris. At that time, the British Green Party's sole member of

parliament, Caroline Lucas, was available to record an interview during the COP but I had to meet her at the National Assembly building in the city centre. I was interested to hear Caroline's take on the various moving policy parts because there seemed to be so many facets that were not in synch and without which, the whole meeting appeared to be farcical.

I raced to the National Assembly and walked through perimeters of heavily armed security and police lines to get to the entrance of the Assembly building. My Blue Zone COP pass dangling around my neck appeared to be an effective method for progressing through various boundaries. I was told to go to the entrance and meet Caroline there during a break when we could find a quiet place to film. I got all the way through security and was looking around. With all my equipment, I stood out from the rest of the bureaucrats streaming about. I waited and waited. And waited. Then, I checked my email again and tried to contact the person who had arranged the meeting. A message came back that Caroline was inside but was not sure when she could get out. Deflated, I made my way outside.

I was exhausted. The relentless days of going to bed after midnight and getting up before 6.00 a.m. were beginning to take their toll. Carrying the tripod and the camera bag, I walked away from the bustle of the security forces until I found an archetypal Parisian brasserie. Inside, I ordered a large coffee and slumped into a chair, contemplating the wasted time and the trudge back to Le Bourget. At that moment, my phone started buzzing. It was Caroline. She had snuck out and was waiting in the lobby. Without hesitating, I tipped the scorching coffee down my throat, paid the bill and ran back towards the security lines in front of the Assembly building.

Once inside, a very apologetic Caroline explained she had arrived at 6.00 a.m. having had no sleep. In London the previous night, she had been a panellist on the BBC's *Question Time*, a political current affairs programme broadcast live on UK television. She had departed for Paris directly afterwards and was exhausted. Apology accepted.

I set up the camera and pinned the lapel mic onto Caroline's green blouse. I started by asking her if she thought there was a disconnect between what scientists were telling us we had to do and how policymakers were responding.

'I find it quite extraordinary when you hear policymakers saying, "Well, we just can't stay below 2°C. If we try to do that, it is going to stop our economies growing." There are some scientific realities here. We have a planet of finite resources. We have an atmosphere that can only absorb a certain amount of greenhouse gases without triggering climate catastrophe. The reality we need to be looking at is the scientific reality. Then we work out how we share out, in an equitable way, the emissions that allow us to achieve that goal, really keeping in mind as well that, obviously, the poorest countries need to be given the most emissions room.'

I asked Caroline what she thought of the non-binding nature of the agreement emerging from Paris. She expressed clear disappointment. Had the deal been binding, the US could not have signed it. Was there anything hopeful about it?

'I guess we can take some comfort, perhaps, from something that Obama said in recent days, that maybe certain aspects of the agreement could be binding,' Caroline continued. 'So, for example, he was talking about the so-called ratchet mechanism, the thing that will allow us to review progress every five years. That process of review could be binding. That is something, but obviously the whole process would be so much stronger if it was mandatory.'

This is the unfortunate reality of the UN COP process, that the power of the US is disproportionately so great and yet encumbered and divided by domestic politics. This means that the best we can do at the highest international level is to remove the teeth from the agreement. The ratchet mechanism will rely on constant public pressure, which is itself divided between those who follow the science, those who deny it, and those who don't care.

'I'm fairly confident that if the political will is there, and if there's enough public pressure on politicians, we could get to 100% renewables by 2050,' Caroline went on. 'But we need the political will, and that's what this process is all about.'

I asked Caroline what she thought of UK Prime Minister David Cameron's record on environmental issues.

'I am disappointed that at the same time that David Cameron is using warm words in Paris, back at home he is taking a wrecking ball to green policy. He is slashing support for renewables. He is building in an illegal obligation to maximise the economic recovery of fossil fuels. What could be more perverse than that? We've got a job of work domestically to make sure our leaders are actually delivering on the ground what they might be pledging here in Paris.'

Caroline's views are pragmatic and respectful of what the science tells us we have to do. This contrasts with the interview with Shannon Phillips. Shannon demonstrated to me no clear, science-based risk assessment of continuing to extract oil in Alberta. It was a short-term political agenda. Caroline used as her starting point the context of safeguarding the people of Britain, while acting with sincerity in speaking with our allies at the UN. This demonstrated a deeper comprehension of the severe challenges, while seeking to respond to them with integrity and, importantly, being able to hold on to the votes in her constituency.

It is a question that must continually be asked: why do so many politicians fail to act with integrity in public office? Why do things have to get so bad before the public will exercise their individual and collective power and say, 'No more'? In Paris, there was so much rhetoric and outright nonsense being spouted by politicians and other commentators. The fossil-fuel industry had more lobbyists in the Blue Zone than we, the public, had fighting for our future.

This creeping power of the fossil-fuel companies into these processes was referred to by Professor Steinberger, herself an IPCC author. She said:

> This is an overt fight. This is the fossil-fuel industry, oil companies, trying to take over climate action, trying to enforce climate inaction by taking over key processes. I was part of the IPCC as a lead author, and the oil companies would comment on our text and say, 'Delete this paragraph.'
>
> It's like, 'Do you have a scientific reason why this paragraph should be deleted? Is there any factual thing that's incorrect in it?'
>
> 'No, just delete.'
>
> It's like, well, no then, we don't have to do what they say.

The failure of the process is reflected in the words of duplicitous governments who say, 'We haven't built the new world yet,' or, 'It will cost too much.' The same politicians who are promoted by the status quo to ensure the status quo.

US Secretary of State John Kerry intervenes

On 9 December, two days before the scheduled end of the COP, word was circulating that US Secretary of State John Kerry was going to make an address to the conference from one of the press rooms. There was a torrent of media people flowing in the direction of the designated location. I didn't fancy my chances of getting in through the scrummage.

Turning, I saw Scott waving, so I navigated the chaotic flotsam and jetsam of delegates, dangerously laden with heavy tripods and dragging cables, to get to him. 'Beckwith has some seats in the US pavilion where they are streaming Kerry's speech on large screens,' Scott told me. 'Wanna come?'

'Sure. Thanks!' I said. We crossed the concourse and entered the large hangar full of glossy pavilions. The US had a large, spinning globe, courtesy of NASA, that impressively showed a number of the Earth's key climate systems in simulation. The rest of the large pavilion was set up with many rows of chairs, anticipating the influx. We headed through the growing crowd to the front and found Beckwith. The air was filled with tension. Were we about to get a sneak preview of what the fabled Paris 'ambition' actually looked like?

After being introduced by the US Special Envoy for Climate Change and lead negotiator, Todd Stern, Kerry appeared on the large screen. The very recognisable, smooth American diplomatic timbre of his voice lent imperial pomp to the proceedings. After a round of thanks, he addressed the importance of Paris:

> One hundred and forty heads of government all came to Paris on the same day to make clear their personal commitment to a global agreement. They all know, as we do, that we have reached a critical moment. We're seeing momentum for an agreement that has never before existed. But at the same time, we are seeing first hand the impact of climate-change. The projections many scientists have been making for decades are unfolding before our eyes. And in some cases, they are occurring faster and with greater intensity than initially foretold. So we gather this week in Paris, knowing that the Conference of the Parties, this Conference of the Parties, may be the best chance we have to correct the course of our planet, that we gather to chart a new path, a sustainable path to prevent the worst, most devastating consequences of climate-change from ever happening.

He went on to denounce the people who call it a hoax or who say that climate impacts will not be that bad, before circling back to the present moment.

'We have to act within the next 36 to 48 hours. We need to get the job done.'

He pursued a more optimistic tone for the challenges ahead. 'Happily, we know time has not run out yet!'

He then went on to compliment all the delegates at the COP, the negotiators, the innovators, the communicators, the indigenous peoples and so on. Kerry's inclusive tone circled the entire COP like long arms locking us into an embrace. Then, bringing us collectively to the altar of hope, he invoked the name of his boss. 'I want to reiterate what President Obama said last week, that the United States of America not only recognises our role in creating some of this problem, but we embrace our responsibility to do something about it.'

That was like the needle scraping across the vinyl of my favourite mood music. Political polarisation amid a heightened level of climate denial and soaring profits from fossil-fuel production did not inspire hope that the United States was going to meaningfully cut back. Only a drastic cessation of new fossil-fuel production, largely championed by the Obama administration, would free up a large chunk of the remaining carbon budget, thus aiding developing nations in a just transition to clean energy.

After about forty minutes, we were released from Kerry's spell. Scott stood up, weighing up his response to Kerry's address. He said, 'Good speech,' and then under his breath, 'but I still think we're screwed!' With that, we joined the throng of humanity exiting the US pavilion.

Divergent narratives emerge from Paris
It was evening in Paris. I was scheduled to take the Eurostar train back to London the following morning. It meant I would

not be here for the closing of the COP. Looking around, it appeared to be a foregone conclusion that we were edging towards a deal. Genuine excitement rippled around the halls; it was infectious. To have a deal – of any kind – between the nations of the world would be historic. Framing it as an agreement that would protect the global commons was very dubious. So-called ratchet mechanisms and talk of raising ambition were the caveats and sugar coating on a pill that hinted at being more of a placebo than the actual cure needed.

I gathered my things for the final time. Mike and Lizzie had already left, but I said goodbye to Beckwith and Scott. In a mood of exhaustion and relief, I headed out into the Paris night for one last meal and a few glasses of Brouilly with some Swiss friends, who had arrived from Zurich to celebrate the unfolding triumph.

Remote closure: 'the lifestyle of some, the livelihoods of the millions'

Back in London, I had received a media alert for a summary press-conference responding to the draft text of the COP21 Paris Agreement. The speakers were the deputy director, Chandra Bhushan, and the director general, Sunita Narain, of the highly respected Centre for Science and Environment, a climate-research centre based in New Delhi, India. Chandra Bhushan began:

'Good evening. My initial comment is that we are actually looking at a weak, inequitable and unambitious deal in Paris. Some of the most important components, including mitigation commitments and financial commitments, are not likely to be legally binding. We find the language on Loss and Damage weak. This deal promises finance but hesitates to commit.'

It is striking that mitigation is the stated main objective of all parties, especially the high emitters, but when it comes to commitments to taking action, the willingness to commit legally is absent. Both Dan Bodansky and Caroline Lucas alluded to the need for the US to be in the deal, even if that meant it was non-binding. Chandra highlighted that whatever deal was reached, the real focus had to be on how the remaining carbon budget would be equitably divided up. This means that countries that have developed their modern way of life by having plentiful access to fossil fuels should reduce consumption and allow the poorer, least developed and developing nations to have access to the remaining carbon budget, aiming to stay below the global 1.5°C mean temperature rise. Chandra assessed whether this requirement was in the deal:

What we are looking at in this draft text is a complete dilution of commitment of developed countries and shifting of burden to the developing countries. For the first time, there will not be a collective target for developed countries on mitigation. There will not be individual targets for developed countries on mitigation. Please remember, from the Kyoto regime, there was a collective target and an individual target. Nothing of that sort is going to remain from 2020 onwards. There is also a lack of a clear time period in which developed countries are going to take collective and individual action. The fact is, the fight here is about the lifestyle of some, to the livelihoods of the millions, and we are really sad to see that in the current text, there is not even a mention of fair allocation of carbon budget. I will now ask Sunita to discuss why the term 'fair allocation of carbon budget' is key for us getting an ambitious and equitable deal at Paris.

In 1991, Sunita Narain had published a report titled *Global Warming in an Unequal World*, dealing with the inequality of assigning carbon emissions like for like in nations of different economic and industrial outputs. For example, tonnes of carbon emitted by people living in relative luxury, flying on multiple holidays or eating red meat every day are not the same as someone in a poorer country emitting a tonne of carbon required to survive.

The result was the insertion into the Kyoto Protocol of 'common but differentiated responsibility and respective capabilities'. This means that developed, wealthier nations should face a heavier burden of emissions reduction because of their legacy of extensive use of fossil fuels. Kyoto was a binding treaty and, although 191 countries signed up to it, the United States refused to sign. Sunita continued:

> We know that the question about cutting emissions is really about sharing the limited carbon budget that remains between now and 2100. We know very clearly that the aggregate emissions which have been promised through the INDCs will not keep the world below the 2°C target. So the entire issue now is, how will you ratchet up so that you have ambition in the deal? And the question of ratcheting up is, who will cut emissions and by how much? The United States, which I particularly want to point out here because it is a country which has come out with all flags flying in this COP, saying that it is the climate-change leader, is one country which has appropriated 21% of the carbon budget between 1850 to 2011. And that is the past. The question is, what do we do in the future? We know the target the world is talking about. If it is 2°C, then you have something like one thousand gigatons [billion tons] of carbon dioxide left between now and 2100. If it is 2°C!

As Kevin Anderson had stated, the carbon budgets provide a clear lens to quantify the action that is (not) being taken to counter the heating climate. They also provide a limited space for each country to reduce its carbon emissions in the future. Adding up all the INDCs allowed the Potsdam Institute for Climate Impact Research (PIK) to calculate by how much we were overshooting the budget for 1.5°C and even 2°C, as Saleemul Huq had said earlier. We were en route to 2.7°C.

The trouble is, as each year passes, the budget is rapidly disappearing. This means the scientific feasibility of staying beneath 1.5°C is diminishing. Sunita continued:

> If it is 1.5°C, the carbon budget shrinks to where you have anywhere between 400 to 550 gigatons left [in 2015]. So the question is, how will you share the carbon budget between the past and the future? If you look at the INDCs of the same countries that I have talked about, we have calculated what would be the appropriation of the remaining carbon budget between now and 2030. If you take the INDCs that have been put forward by the United States, then 21% has been appropriated in the past, and the lack of ambition of the US means it will appropriate another 8 to 10% between now and 2030. And what we at CSE have been saying, again and again, is that if you put together the INDCs of all the countries, then 80% of the remaining budget for 2°C finishes by 2030. The chart put forward by the UNFCCC itself accepts that by 2030 the bulk of the remaining carbon budget will be finished.

Sunita also outlined the near impossibility of holding to 1.5°C because to do this in a fair and equitable way would mean high-emitting nations accepting negative growth. Rapid emissions reduction by high-emitting nations, or *mitigation*, is the one climate-change strategy that scientists prescribe and politicians ignore. 'We want the world to stay at 1.5°C,' said Sunita, 'because, coming from India, we are already seeing the worst impacts of climate-change. We know the pain of extreme weather events. But if we want a reference to 1.5°C then we want an absolutely clear reference to sharing the available carbon budget; otherwise, let us be very clear, we are not operationalizing equity. We are creating a climate-change regime which is built on intolerance of the voice of the powerless, which actually furthers climate-change apartheid.'

On 12 December 2015, 195 nation states and the European Union agreed on the final wording of the Paris Agreement. Figueres had got her deal. The COP president, Laurent Fabius, pounded the gavel, signalling that COP21 had reached its conclusion. The crowd in the room, including all those on the panel, erupted like fans of a major football team scoring the winning goal in a championship. Al Gore leapt up, cheering the home team to victory, as others hugged and cried tears of joy.

The official boundary for warming in the Paris Agreement was 2°C, with a commitment to hold as close as possible to 1.5°C. This figure of 1.5°C was nearly omitted altogether, but the vulnerable nations pushed to have it included. Even if it was acknowledged as hard to achieve, it was hoped it would create downward momentum for emissions reduction.

The triumph was in the diplomatic achievement of getting the countries of the world to agree loosely to an intention. Every person who accepted the science of climate-change believed something remarkable had happened in the timeline of humanity. As far as solving the climate problem was concerned, that was a task for another COP, another diplomat, another generation.

3

COP22, Marrakech, 2016

The flight path from London to Marrakech is practically a straight line. As the plane hurtled southwards over what appeared to be an entirely unpopulated expanse of Spanish landscape, there was a curious sense of unease occupying my thoughts.

This was because on the previous day, the United States had elected Donald Trump to be their next president.

The previous COP had ended with an overall feeling that something had been achieved but that it was tenuous and required nations to both honour their commitments and increase their 'ambition'. While campaigning, Trump had declared that he would withdraw from the Paris Agreement as soon as he took office. COP22 was already underway when the election result was called, and I could not imagine what the atmosphere was going to be like.

Images of the descending gavel and the jubilant politicians leaping into the air in Paris were slowed down in my mind, melting like old film against a burning projector bulb. How things had changed. Christiana Figueres had stepped down and been replaced by Patricia Espinosa, Mexico's former Secretary of Foreign Affairs. I knew little about Espinosa, but was curious to see how the incoming Executive Secretary of the UNFCCC would address this very obvious politico-climatic disaster.

We crossed down over Seville and tracked along the Atlantic coast of northern Morocco, turning inland after the capital, Rabat. The aerial view of the land below was strikingly green, divided into small parcels being utilised for agriculture. It appeared much more fertile than the southern half of Spain over which we had earlier flown. This green soon gave way to a creeping cumin-coloured rockiness, with splashes of turmeric yellow as the edges of the Sahara crept northwards. Soon, the defining outlines of Marrakech emerged to meet us.

Stepping off the plane, the bright, dusty heat of North Africa replaced the cool chill of a wintry London. A whiff of aviation gasoline in the warm breeze caught my nose while we gathered to be processed by immigration.

The Moroccan airport staff greeted us with smiles, which quickly turned to speedy efficiency as they rounded up the COP delegates and led us outside to waiting buses. I must admit that I had expected chaos, yet despite the incredibly noisy terminal, filled with shouting, alarms and various beeps, we moved with ease through it all.

It took about forty minutes to get us to the edge of the famous Jamaa el-Fna Square, where I pulled my vast suitcase from the storage compartment of the bus and dragged it into the crowded, pulsating hub of the Medina.

It was good to be back here in this mad place. Gazing around, I felt an urge to be lounging with a sweet mint tea on the roof terraces of the cafes surrounding the square. I followed my printed directions into the warren of alleys that run down the side of the huge souk, towards the concealed riad.

As I fended off a relentless onslaught of market traders with their wares – slippers, jewels, pottery, perfume and more – the aromatic fug of fresh tanned leather, spices and floral fragrances filled my nostrils and induced a charge of energy and excitement that had so far eluded me.

Eventually, I arrived at a small alleyway shooting off to the left and followed it to its conclusion, where an ornate carved wooden door stood before me in a windowless dead end.

I felt myself being examined through a spy hole in the carved edifice before the door opened fully and a beaming young Moroccan man came out and attempted to lift my case. This was ill-advised; the case contained two tripods, cameras, a litre of gin and even more of tonic, shoes, clothes and plenty of cables. After he reluctantly gave up, I took the strain and lugged it to the reception.

Entering the riad, things couldn't have been more different. The stillness and calm pervaded the negative space between visitor and surroundings. I ascended the staircase with my bag and arrived at the first landing. As is typical with riads, each floor faces inwards, with a landing that runs along each side facing the central courtyard area. The intricately carved stone walkways and balustrades facing the central lightwell were typically ornate and beautiful. In our central area, there were some tables and a small, mosaic-tiled swimming pool. The landing had windows from each room opening onto it; there were no external windows.

Given the time, I headed into the COP to register and get set up before heading back for a relaxing evening. Work would begin in earnest the following day.

Back in Jamaa el-Fna, the din was unmistakable for anywhere else: a constant clatter of chants, drums, a mixed tapestry of humanity thronging around. I bought a fresh pomegranate juice as the sun blazed down. The essential tartness of the juice, lifted by sweet notes, refreshed my palate. For the esoterically minded, this was a nest of dubious pleasures: palm readings, henna tattoos, snake-charming, infant Barbary macaque monkeys trotted out in chains for tourist photos, belly dancing, storytelling, food stalls – all on offer to the same subliminal beat.

Looking around, I noted the entrance to the square where I had arrived. There was signage for COP22 with a scattering of delegates, easily recognisable with their Blue Zone passes hanging around their necks. I headed over and boarded a shuttle bus. All I had to do now was sit in thick traffic as the bus driver patiently edged along at walking pace. It was more or less a straight road to the 'COP village' and the nearest corner of the vast conference complex was not far off. The actual entrance, though, was at the furthest end, and that *was* quite far away.

The COP is a travelling tented circus of its own. Once inside, you see the familiar interior where everything is a vague copy of the previous one, except for some regional branding tacked on and perhaps a slight rearrangement of the temporary structures.

I got my badge and walked through security into an open concourse beneath the gorgeous blue November sky. Walking with my rucksack and tripod towards the media centre, of all the thousands of people I might have met, it happened to be Scott walking towards me. In the wake of Paris, we had not stayed in close contact and I had decided to focus on my own work at this and subsequent events, rather than spend much time with others. He approached now with a dejected air.

'Hi Scott, how are you?' I asked.

'Hey, I'm good, thanks.' This answer did not match his demeanour.

'Yeah, well,' I continued. 'How long have you been here?'

'Since last week.' He shook his head with a glum, downturned mouth. 'But the US election has really hit me hard. No one knows what is going on. I don't even know if the COP will reach the end.'

'Wow, yeah, it is big news,' I agreed. 'Trump has said he will withdraw from the Paris Agreement.'

'Yeah, but we're screwed *anyway*, so I guess it doesn't matter.' His resigned expression for encapsulating environmental doom had an air of tranquillity.

'Well, you never know. Maybe Trump *is* the wake-up call. Clinton would have been business as usual, no?'

'You think so? But the guy is a stupid asshole.'

'We can only hope. I am going to get on. I'll see you around,' I said, moving onwards, along the paved concourse.

The industrial sprawl of the Paris COP was replaced in Marrakech by a dustier desert spread. The large press-conference rooms were on the left side of the long, straight, open-air concourse, followed by a sprawling media centre beyond them at the end. Looking back towards the entrance from the media centre, on the opposite side in separate, much larger temporary buildings were the plenary rooms, the delegation pavilions, and the side-event area, as well as the smaller NGOs lumped in with other exhibition stands.

Inside the media centre, I took a moment to scan through social media, looking for reactions within the COP. A statement from Executive Secretary Espinosa sprung to the top, reading diplomatically: 'I would like to congratulate Mr Donald Trump on his election as President of the United States, and we look forward to engaging with his administration to take the climate action agenda forward for the benefit of the peoples of the globe.'

None of this boded well for COP22, officially tasked with the real purpose of beginning the process of implementing the Paris commitments. On top of that, the wealthier, high-emitting nations were meant to be agreeing on a package that would concretise the long-before-agreed $100 billion a year. This was to be paid to the poorer, more vulnerable nations that were already dealing with climate impacts, despite not having caused them. It was hard to see how any of this could progress.

The next quote that appeared was by the German climate scientist, Hans Joachim Schellnhuber, then director of the Potsdam Institute for Climate Impact Research, who stated: 'President-elect Donald Trump's stance on global warming is well known. Science cannot expect any positive climate action from him. The world has now to move forward without the US on the road towards climate-risk mitigation and clean-technology innovation.'

Schellnhuber's comment was devoid of spin and cut through the diplomatic niceties that were filling the official channels. Whatever the criticism might have been regarding the outcome of the Paris COP, the early signals in Marrakech were looking brutal.

After getting my work station set up in the media centre and storing my equipment in my locker, I walked past the side-event area, towards the exit. This was generally a place out of my area of interest, but the noise that was emanating from it had piqued my curiosity. It was packed out with an international audience of all ages, hanging on the words of a very young American guy on stage who was wearing a pinstripe suit. His face was red and he appeared enraged, screaming into the microphone that political failure (apparently referring to the US election) would not be tolerated. He called for 'climate justice', a 'rapid, equitable transition to clean energy', and an end to 'all planned geoengineering research'.

The last of the three was unexpected. Geoengineering research to date was insignificant in contrast to the scale of the challenges that the researchers involved proposed to confront. In terms of planning, any research was apparently in the hands of social scientists, discussing hypothetical governance issues. Despite this, it was his evident anger that made the initial impression on me.

I turned to head in the direction of the COP exit, noticing the sun sinking. The last light was casting long shadows, giving the place an elegant film-noir feel. It was these long shadows that interrupted my line of vision on the pathway on the left side. I turned my head to see a group of people standing around a small dais and a woman announcing in a comedic voice – that was like a town crier combined with chiding school teacher – the 'Fossil of the Day Award'. It is a small event I had also witnessed in Paris that gives the award daily to the nation that is doing the least to contribute to the overriding goal of the COP. Given that the US elections had just surgically removed the COP's heart – without anaesthetic and whilst failing to provide a transplant – this display created a sense that we were unwittingly inside the body of a corpse, whilst undertakers outside disposed of its flesh.

After a long trek back to the riad, I was cheered to discover a large roof terrace with comfortable seats. I ambled up with my bottle of warm gin and some tonic and took a seat with a view across the rooftops. I sipped my drink and listened to the melodic call of the muezzin, the servant of the mosque who calls the faithful to prayer five times a day.

Suitably inoculated from the rising terror of background events, I walked back to Jamaa el-Fna to visit the evening food stalls that bustled with Moroccans and tourists alike. Aromas of fried fish and spices filled the air, cleansing the mind and arousing my appetite.

COP autopsy

The next day, I rose early and headed into the COP. Inside, the mood seemed to be increasingly bleak. Most of the younger civil-society delegates I saw looked grief-stricken, others angry, or just sullen. Walking along the concourse on the way back from recording an interview, I saw a fast-moving scrummage of people with Espinosa in the middle. Tight-lipped and moving at speed, possibly to avoid press entrapment, there was none of the usual swagger of officialdom. Instead, they passed as fleeting apparitions.

I conducted some interviews with people from various NGOs – the European Investment Bank and others – but the reason for being here had evaporated. COP22 was over before it had begun, and it leached away the outgoing optimism of the Figueres era of global diplomacy.

Throughout these negotiations, it often ends up being the United States that scuppers hope. It builds them too, but ultimately, the American colossus of consumerism and carbon pollution, hindered by a blindness to its own appetites, sets the currently low probability of survival for our own and many other species. Speaking of the COP process, New Zealander David Tong, leader of the Aotearoa Youth Leadership Institute's delegation at COP22, was quoted in the *Pacific Standard* magazine as saying: 'We learned in Copenhagen, if not before, that the United Nations will not save the world. We should damn well know by now that governments will not save the world. It's people that save the world. Governments just tend to get in the way.'

Something in this comment hints at a key observation that recurs over the course of these conferences and which Naomi Klein alluded to in Paris. We cannot outsource our agency when the stakes are so high. Marrakech was at political *absolute-zero* for global climate action, but what if a shift in the possession of agency could be moved away from such a dysfunctional political elite? Echoing Scott's environmental doom-speak, Tong's comment ended with, 'Even if we fail, and we get a 3.5 or 4.5 degree world, I can't think of a bunch of better young people to fail to save the world with.'

The mood of political failure was engulfing us. *Actual failure*, in the broader arc of our collective actions, was yet to be determined.

Events had moved so fast that the US negotiators were still in Marrakech, rendered impotent by events. The outgoing Obama administration, like the rest of us, appeared in a state of shock. The upper echelons of the US administration were represented at the COP by Secretary of State John Kerry. He had spoken in great optimistic tones in Paris, as the momentum had grown towards a globally inclusive and diplomatic outcome.

Kerry stood before us in an overheated temporary press-conference room, so packed with journalists that the building wobbled portentously throughout the speech. He was here to deliver a very different set of crafted words to those upbeat ones he gave in Paris. His demeanour was now darker, the timbre of his voice traversing a line between hope and despair as he said:

> In no uncertain terms, the question now is not whether we will transition to a clean energy economy. The question now is whether we are going to have the will to get this job done. Whether we will transition in time to do what we have to do to prevent catastrophic damage.
>
> Ladies and gentlemen, I'm not a Cassandra, as you can tell from what I've said. But I'm a realist. Time is not on our side. The world is already changing at an increasingly alarming rate with increasingly alarming consequences. The last time that Morocco hosted the COP was in 2001, and the intervening fifteen years have been among the sixteen hottest years in recorded history. 2016 is going to be the warmest year of all; every month so far has broken the record and this year will contribute its record-breaking heat to the hottest decade in recorded history, which was, by the way, preceded

by the second hottest decade, which was preceded by the third hottest decade.

The largest emitter of carbon pollution in this period of fifteen years between the Moroccan COPs had been the United States. As the wealthiest nation on Earth, the US has consistently given the rest of the world, including the most vulnerable nations, a shrugging reply that 'our way of life is not for negotiating'. It harks back to the end of COP21, when Dr Sunita Narain specifically called out the US for their persistent greenwashing, saying, 'The United States. It is one country which has appropriated twenty-one per cent of the carbon budget between 1850 to 2011.'

Kerry continued:

At some point, even the strongest sceptic has to acknowledge that something disturbing is happening. We have seen record-breaking droughts everywhere from India to Brazil to the West Coast of the United States. Storms that used to happen once every 500 years are becoming relatively normal. In recent years, an average of 22.5 million people have been displaced by extreme weather events annually. We never saw that in the twentieth century. Communities in island states, like Fiji, have already been forced to take steps to relocate permanently because the places they have called home for generations are now uninhabitable. And there are many, many more who know it's only a matter of time before rising oceans inundate their cities. I know this is a lot for anyone to process.

Now, in anticipation of his successor who would serve under President Donald Trump, Kerry offered some words of guidance. Given the disproportionate contribution of greenhouse gases to the atmosphere that the US was continuing to make, the despair in the following script is as much for the failures left behind as it is for what lay ahead:

> For those in power, in all parts of the world, including my own, who may be confronted with decisions about which road to take at this critical juncture, I ask you: examine closely what it is that has persuaded popes, presidents and prime ministers – leaders around the world – to take on the responsibility of responding to this threat.
>
> Speak with the military leaders who view climate-change as a global security concern, as a threat multiplier. Ask farmers and fishermen about the impact of a dramatic change in weather patterns on their current ability to make a living and to support their families.
>
> Above all, consult with the scientists who have dedicated their entire lives to expanding their understanding of this challenge and whose work will be in vain unless we sound the alarm loud enough for everyone to hear.
>
> Anyone who has these conversations, who takes the time to learn from these experts, who gets the full picture of what we are facing, I believe, can come to only one legitimate decision: that is to act boldly on climate-change and encourage others to do the same.
>
> Now, I want to acknowledge that since this COP started, obviously, an election took place in my country. I know it has left some here and elsewhere feeling uncertain about the future. I obviously understand that uncertainty, and I can't stand here and speculate about what policies our president-elect will pursue.
>
> So we have to continue this fight, my friends. We have to continue to accelerate the global trend to a clean energy economy and we have to continue to hold one another accountable for the choices that our nations make.

Everyone in the room must have accepted the sentiment of the words, but many would also have felt the disappointment of the Obama years, compounded by the failure of the US political system as a whole to wield its great power in a far more globally collective and, at home, non-partisan way.

He stepped away to thundering but sullen applause. On the way out, I bumped into Professor Jason Box. We agreed to meet later with Professor Hugh Hunt on the roof terrace of the riad for a preprandial drink and dinner. Over a Moroccan tagine, washed down with the Moroccan red wine Cuvée du Président Cabernet, we talked about dystopian politics, Greenland's melting ice sheet and the likelihood of storing a trillion tonnes of carbon from the atmosphere.

A final salvo from Marrakech

Children and future generations are often cited in climate discourse when emphasising the injustice of not taking appropriate action. In Paris, I paid little attention to the youth delegates, but on the last day at COP22, Hugh and I walked in on a press-conference being given by YOUNGO, the UN Youth Delegates.

The YOUNGO panel comprised a broad spread of international youth, voicing their concerns about the lack of progress in Marrakech. They made clear their frustration at being kept out of the talks that decided their future. After the press-conference, we spoke to one presenter, Jeanne Martin, who gave some interesting insights into what all this looked like from the youth perspective. Jeanne was French, but living in London and here with the UK Youth Climate Coalition. Ms Martin explained:

> Looking at the science, the logic is that we need to take action now, and the parties keep on delaying action. They keep on

postponing sessions. So, for example, they were supposed to talk about the adaptation fund, about implementation mechanisms for NDCs [Nationally Determined Contributions], agricultural issues. Everything has been postponed either to COP23 in Bonn in 2017, or 2018.

There isn't a sense of urgency that should drive the negotiations forward, and that's a shame. That is why we are here: to re-inject a sense of urgency into the negotiations. At least, that is what we are trying to do.

In all that Jeanne said, there was no reference to the political events in the US. Her focus was rationally on the scientific reality. Hugh asked if she thought the policymakers knew they were failing to deliver at these COPs. 'Well, I believe that, yes, some of them must know,' Jeanne replied, 'because a quick look at the science makes it clear that we need to act now. Parties have committed last year to aim at maintaining global temperature rise to 1.5 degrees.'

Professor Hugh Hunt said, 'It isn't possible, is it, 1.5 degrees?'

But Jeanne Martin answered: 'Carbon Brief showed that it was possible that we would have a sixty-six per cent chance of meeting that target if we took action now, because we only have five years left to peak emissions and then remain under 1.5 degrees.'

It is worth interjecting to emphasise that in 2016, a 66% chance of holding the global mean temperature rise of the planet to 1.5°C above the pre-industrial period, with five years to peak global emissions, is again a reference to the numbers that are meant to be informing the COP outcome. The lack of progress to date and the recent election result motivated Hugh's scepticism that this could be achieved. In 2023, it was calculated that the remaining carbon budget (the quantity of carbon pollution we can collectively emit) that would give us a 60% chance of staying within the 1.5°C danger threshold is being consumed at 1% per month. A clear warning that every month of burning unrestricted fossil fuels reduces our chances of holding on to a safe climate.

Hugh continued by asking Jeanne if she thought the politicians could agree on anything that would bring the emissions down. Jeanne replied, 'I hope so. Given the political reality, I don't think it's gonna happen, but I don't think that's a reason not to push for it, not to spend all of our energy lobbying for it and doing everything we can. I think it is easy to get depressed when you work in the climate-change arena because it is such a complex, complex problem. It is just easy to feel powerless. So it is good to meet young people who are taking strong action in their country.'

The depression that Jeanne speaks of is a growing reality among young people. It is creating a generation of angry men and women. I couldn't help but feel empathy for these youngsters who are confronting the world's toughest issues. They have no other option but to look into the void for new avenues to shape their own futures.

The YOUNGO press-conference had taken place at the very end of the COP, probably after many top-level UN and other officials had flown home. Yet, to us, this exchange had completed the trip. Hugh and I asked Jeanne to participate in the newly formed Cambridge Climate Lecture Series that was set up with other colleagues in memory of the late British climate scientist, Professor Sir David Mackay. We were organising an end-of-series panel in Cambridge for early the following year. To have such a thoughtful young person, with experience of the COPs, would be a great asset.

Jeanne tentatively accepted and went on to make a powerful contribution to our end-of-series panel in Cambridge in March 2017.

4

Interlude, Christiana Figueres, 2017

In April 2017, I received an invitation to go to the London Google HQ to interview former Executive Secretary of the UNFCCC, Christiana Figueres, at the launch of a new initiative called Mission2020.

I walked in glorious sunshine across central London to get to Google's HQ, just north of Covent Garden. The chilly wind crossing Waterloo Bridge was bearable but in the shade of the buildings, I braced myself until the next spell of warmth. Inside the Google building, there was an unfamiliar yet familiar feeling, a kind of brand déjà vu. The auditorium was arranged with rows of about two hundred chairs, set before a multi-screened stage. Across the screens in bright, bold lettering were the slogans '2020, The Climate Turning Point', 'Necessary, Desirable, Achievable', and '2020 Don't Be Late'. With the rest of the large room in low light, it was the aroma of fresh-baked pastries and coffee that served as my compass.

Christiana arrived with a small entourage, herself in the centre, engaged in enthusiastic chatter. Pouring another cup of coffee and wiping away the evidence of crumbs, I returned to where I had set up my camera and waited until a jolly Christiana sat down in front of me. At close quarters, she appeared good-humoured and elegant, wearing a pale-pink shawl-collared jacket and a pearl necklace.

Mission2020 focused on six key milestone targets: energy, infrastructure, transport, land use, industry and finance. The aim of the initiative was to cajole each industrial sector to align their practices within the framework of the Sustainable Development Goals (SDGs) by 2020, ensuring they were then on track to meet the next established milestone of 2030.

It was 10 April 2017, and the repeated intention of the Trump administration to withdraw from the Paris Agreement was an affront to Figueres's COP21 legacy of multilateral wheeler-dealing.

Christiana immediately teased me about my 'cheat sheet' of questions. So I teased her back, pointing to the Mission2020 logo on the large screen and asking, 'As noble an effort as it is, shouldn't we be calling this *Mission Impossible*?'

Christiana replied, 'I don't think Mission2020 is completely impossible because the fact is, we just have confirmation from different independent sources that for three years in a row we have flattened out greenhouse-gas emissions and we have an increasing GDP. So we could already be beginning to decouple greenhouse-gas emissions from GDP. The fact is, we are already walking in the right direction. Now what we are trying to do is just increase the pace and the scale. We have to be at a decarbonised economy by 2050.'

I followed up by asking if the Trump administration's anti-climate rhetoric had dissipated the momentum from Paris. She said, 'Well, I think the international political arena is going to have to make their adjustments if the United States administration decides to do something differently under the Paris Agreement, which we don't know yet, but politics is one thing and the real economy is another. The real economy has not stopped. Neither in the United States nor anywhere else, because they do know that this is economically beneficial.

So, I am not concerned about reversing the direction. I'm concerned about keeping the direction and increasing the speed.'

The post-Paris rhetoric was on the coming together of a trinity comprising civil society, politics and business. With the British public consumed by Brexit and the American political situation verging on chaos, I asked if we were expecting too much of business to carry the baton all the way to the finish line. Figueres said, 'No, I don't think we're expecting too much of business. Business is not doing this to save the planet, believe me! Business is doing this because it is good business, because there are safer investments, because there are less risky investments and because the prices are simply going down at such an exponential rate that it just makes more sense to invest in clean technologies than in technologies of the past century. Business is doing this because it is good for the bottom line!'

Emissions from transport were highlighted in the Mission 2020 report as needing to come down. Aviation is a huge part of emissions from transport. I asked, if we are to achieve these goals, should we be flying less? Christiana said:

> The fact is, you cannot exempt any sector of the economy from these efforts. So you can't say, 'We're not going to fly because aviation is too high emitting.' That is the wrong approach. The approach is aviation and maritime and land transport. All three of them have to come down in their emissions.
>
> And it's very interesting that in the last two weeks we have had an announcement from a very small start-up, as well as from Siemens, that aviation is moving toward electrification. They foresee that ten years from now, they will have aeroplanes that are fully electric with clean energy and they will have a thousand kilometre range. So you already have in a very short time-span flights that can be clean – certainly the short flights,

and then we have to look at the longer ones. But it is not a question of changing one mobility for another. For the time being, if you want to be responsible, yes, definitely go for the mobility with the lowest emissions, but that cannot exempt any sector. Every sector has to bring down its emissions, and aviation is coming!

To suggest the aviation industry is going to transition to electric planes within the time frame of the next decade, or even two, seemed to me highly improbable and much more like aviation greenwash; in other words, 'Keep flying and we'll switch to electric in the future with minimal interruption of service.' It felt hugely optimistic to be citing these reports as tangible advances.

The aviation industry benefits from being subsidised by not paying duty on aviation fuel, as well as in many other areas of their business. Prices of flights are kept artificially low because airlines don't pay the true cost of the environmental damage they create. Choosing 'to be responsible' by using alternatives, such as trains, is often not so easy because of under-investment and the reality that the tickets for the same journeys by alternative means of travel cost a lot more than flying, despite their far lower carbon emissions. Flying for a holiday once a year differs greatly from those who are airborne every week or every month – in fact, I have been in the latter group myself. Yes, aviation has to reduce its emissions but the only way to do this in the time frame we have is to fly a lot less and de-grow the industry significantly, while making low-carbon alternatives, such as trains, much more widely available and affordable.

In my final question, I asked Christiana what the catalyst was that would direct the flows of investment capital towards meeting the Mission2020 goals. She replied, 'I think what directs the flow of capital is actually cost, and renewable energy is, every day, just more and more cost competitive. You can see the flow of capital going into renewables and not into oil. It is just very evident.'

It is true that the price of renewables is coming down significantly and investments are rising steeply, but so far total global emissions have remained stubbornly high. The fossil-fuel companies have undermined our efforts at every turn to accelerate towards a cleaner and more efficient future that would provide energy security and a cleaner, healthier planet. In 2017, fossil-fuel subsidies alone were in excess of 400 billion US dollars. In 2022, they soared to 1.1 trillion. This is a vast amount of money that could have been invested in transforming our societies. The longer we put off these meaningful conversations about the future, the fewer choices and opportunities we will have to create equitable and, in terms of resilience, secure outcomes.

Mission2020 was a clear effort to move beyond policy and put pressure on industry. This was a response to the political climate in 2017, but, nonetheless, it seemed to run counter to the celebrations of COP21 in Paris, where politics was central to the narrative of climate action. I was reminded of what Caroline Lucas MP had said in Paris about business needing clear policy. We were in London, and since Paris, Prime Minister David Cameron had resigned in the wake of the Brexit referendum. His replacement, Theresa May, had made a lot of noise promoting a green agenda, but the bitterly divided political climate, not to mention the growing number of climate-change-denying MPs in her Conservative Party, meant nothing translated into a policy that would accelerate emissions reduction.

Following the interview with Christiana, I retired to the back of the room as upbeat guests poured in. Happily, the buffet had been upgraded, offering both red wine and croissants. I offered libations for the success of Mission2020, but had my doubts.

5

COP23, BonnFiji, Germany, 2017

I was not keen to attend COP23. After Marrakech, the entire COP reason for being felt compromised by gaming politicians and a disinterested public. I will credit Professor Hugh Hunt for encouraging me to attend. As part of the Cambridge Climate Lecture Series, he was proposing we take some students to Bonn to witness how they perceived the safeguarding of their future to be progressing.

COP23 was officially to be hosted by Fiji, one of the small island developing nations severely at risk from +1.5°C heating and a broad range of associated impacts, including sea level rise, more intense cyclones, droughts, coral reef degradation and human health impacts. Tens of thousands of delegates flying in and out of Fiji would run counter to what could be considered a sustainable intake of visitors. The UN HQ in Bonn was therefore designated as the venue for COP23. Thus, it was branded BonnFiji.

I arrived to find a cold, wet and blustery Bonn. My lodgings were a ten-minute walk away from the conference centre. We had just crossed the line from autumn to winter and the naked trees set a sombre, bedraggled scene. Inside the COP, I went through the usual rigamarole of getting a badge, finding a workspace and locker, and getting my bearings. One of the first people I encountered as I walked around was Professor Kevin Anderson. We agreed to catch up in due course. Kevin has had a no-fly policy for many years and so was not in Marrakech at the *non*-COP.

The conference was laid out more unusually, with negotiations and media centre on one side and the pavilions, NGOs and action spaces on the other. The space between was a twenty-minute walk, but e-bikes were laid on, if you could locate one.

I soon bumped into Scott, who was doing his Climate Show. He had a little more of his angsty self on display, welcoming me with a scowl. His Climate Shows were taking place in the COP press-conference rooms, which, by definition, meant they were for media consumption. However, Scott went to great pains to explain to me that his sessions were not operating under the same rules, telling me that, 'I don't want you filming in my shows.' I pointed out that he had no copyright control in a room designed for media. As I walked away, I could feel his perturbed eyes burrowing into my back.

What made this COP special was something different from the UNFCCC furniture, which reinforced a familiar, foreboding ambience. Escaping the plenary halls and media centre, heading across to the Bonn Zone, located at the far end, the room's energy changed. It was abuzz with people, predominantly young folk.

I am not sure if it was the offspring of the YOUNGO crowd in Marrakech, or the simmering masses of a generation provoked into taking a stand. Whichever it was, there was a diverse flood of young people, intermingled with indigenous peoples from around the world, who were sitting in groups, conversing, conspiring and, importantly, rising.

This atmosphere continued to grow and grow. I wasn't the only one to notice the evolving of these spaces. With the United States withdrawing their presence at the federal level under the Trump presidency, a new US arena had popped up in a prominent position. Saleemul Huq described it when I caught up with him, explaining that: 'It's a non-governmental enclosure. It used to be a government one. So that's good. It illustrates, in my view, the shift in emphasis and responsibility away from governments to negotiate text, to people to do things to implement the elements of the Paris Agreement that are relevant to them. Mitigators and emitters need to reduce emissions, and people who are vulnerable need to adapt.'

If we were expecting a void to be left by the US, here is Saleem's description of the space that before was unseen:

> One of the things that we are seeing in COP23, a good example [is] we are sitting now where the negotiators are, and they are just milling around, you know, nothing much exciting happening here. Sometimes they sit behind closed doors and we sit on the outside. We don't have any idea what they're talking about. Whereas a few kilometres down the road, they have what they call the Bonn Zone, which is vibrant and pulsating with music and NGOs having meetings and sit-ins and demonstrations. It is far more exciting. And you now have mayors of cities and governors of states, even in the United States, coming in, pledging action.

I could attest to this. On one visit to the Bonn Zone, I watched the former 'governator' Arnold Schwarzenegger come and go, leading a pack of admirers in a jog around the main area. Upon returning, he let out a belly-laugh and did the whole circuit again. Along the same route, the actual serving governor of California, Jerry Brown, stood speaking to reporters, saying he was in Bonn, despite the federal government, to pledge that California would reduce carbon emissions in line with the Paris Agreement.

Despite all this, the negotiations that decided the politics were still down the hall. Hugh arrived in due course with four students from Cambridge. Although unsure if they would get passes, they eventually appeared on the inside, having made it through the semi-porous borders. Scott had reluctantly, given that there was no quid pro quo, agreed to get the youngsters passes.

Professor James Hansen also attended to take part in press conferences and interviews. I recorded one myself with him. It was an echo of previous interviews. I published it anyway, but it was a signifier to me that nothing was moving. Even the movement wasn't moving.

One refreshing interview was with the former head of the UK Green Party, Natalie Bennett. An Australian by birth, she is one of the most recognisable faces in British green politics, having been given a seat in the House of Lords.

I asked Natalie for her take on the politics here at COP23 and she told me: 'There is a real urgency gap with people talking about things we will do after 2020. Yet the reality is we have to do things in the next two or three years, or certainly, we are not going to get to 1.5°C. I think scientists tend to look at what governments are doing now and say, "Oh, well, you know what, we will see some slow, gradual improvement." What we actually need to see is a radical change. What we need to do is deliver that real absolute political will to get the total change. And 1.5°C is immensely challenging. There's no doubt about that, but it is doable.'

Professor Kevin Anderson presented another dire assessment during an all-star line-up of big names that included former Irish President Mary Robinson, Christiana Figueres, Germany's most eminent scientist, Professor Hans Schellnhuber and Professor Johan Rockström from Sweden.

Kevin called his talk 'Beyond Nebulous Arm Waving: A Look at What a 2°C Strategy Might Be', and here is a taster:

I think a bit of humility, to start with, from people of my generation with grey hair, no hair or dyed hair. In 1990 we had the first IPCC report, and this year global emissions will be sixty per cent higher than they were in 1990. In fact, we had the figures yesterday from the Global Carbon Project that emissions are going up again. They are two per cent higher.

So what we have had from my generation is twenty-seven years of abject failure on climate-change since the first IPCC report. So a bit of humility is important from us.

Also, what have we done in that period? I would provocatively say we have had a litany of scams. We have had offsetting, getting the poor to pay by dieting for us. We have had the Clean Development Mechanism (CDM), making offsetting legal by governments. We have had emissions trading schemes, with a very low price for carbon so it makes no difference. We have now got negative emissions technologies (NETs) in the pipeline, and then we are planning for geoengineering. What we haven't tried in twenty-seven years is mitigation – actually reducing emissions!

Kevin quoted Pope Francis's encyclical, *Laudato Si'*: 'The alliance of technology and science ends up sidelining anything unrelated to its immediate needs, whereas any genuine attempt to introduce change is viewed as a nuisance based on romantic illusions!'

His presentation highlighted the failures of those in charge of dealing with the climate issue effectively. In turning attention to the climate-change community itself, including those present in the room, he interpreted the actions to date in this way:

In developing 2°C emissions scenarios, we have fine-tuned our analysis to fit with political and economic sensibilities. Universities and NGOs have been co-opted by near-term power. We want to be at Davos, we want to be at the meetings with the great and the good in our society. We fear questioning the socio-economic paradigm; that is much more important than physics, apparently. Engineers, like myself, have a naïve focus on particular pet technologies. Whether it is nuclear, wind or solar, it is always on the supply side.

He goes on to highlight how the mitigation pathways presented by the IPCC embed 'negative emissions technologies' in the forecast reductions, so that the challenge doesn't appear so staggering. The problem is that the technologies have not been developed. Aside from planting trees, which have limiting factors in their overall impact, the novel technologies that are meant to scale up to draw down billions of tonnes per year have not progressed to the real world. In 2023, they comprise a fraction of 1% of actual global negative emissions. By continually inserting them into the integrated assessment models, we are creating a very misleading approach to critical policymaking and implementation.

Kevin continued, saying, 'We are doing what the pope warns us about. We are relying on a utopian alliance of technology and economics. This is what informs all governments around the world; we are relying on very speculative negative emissions technologies to suck carbon dioxide out of the atmosphere in the future. So the buck is passed on from my generation to the next generation because my generation can't be bothered to make real changes, and we are passing it on, beyond the end of the century.'

He ended by repeating what he had said in Paris regarding the wealthiest 10% of society creating 50% of the global carbon emissions. Cutting the emissions of that 10% to the same level as the average European would reduce global emissions by a third. Emphasising the point that the 10% were not a vague subsection of society, he said, 'Just look around us here at COP and listen to the discussions you will get. People meeting in Mexico or Vienna. How are they getting there? They are flying there, back and forth around the world. Climate-change scientists, academics and NGOs, flying around the world to discuss with each other how to reduce emissions.' Closing with one last glance around the room, he added, 'And the climate glitterati are generally in business class!'

As with Jim Hansen, the actual content of these presentations is not different. What *was* changing was the mood. The climate glitterati were not delivering. New research also attributed the source of global emissions directly to the largest fossil-fuel corporations. It was Saleemul Huq who really spelled out the bitter taste this murderous profiteering was leaving on the palates of the climate-vulnerable nations. He said, 'The time has come to put "the polluter pays" principle into practice, and in this context, we have a set of maybe a few dozen very large multinational fossil-fuel companies who are making billions of dollars of profits by mining the fossil fuels and selling and burning them; causing the Loss and Damage. We know who's causing it. They are companies and they're making profits out of it. So tax them; let's put a Loss and Damage levy on them and let the countries apply that levy and put that into a Loss and Damage fund. Probably raise billions of dollars that way.'

It sounded improbable to me, but I asked Saleem if it was at all realistic. He said, 'We're not going to get it here, but we're going to push for it. Maybe not COP23. Maybe COP24. If not COP24, COP25. But sooner or later they're going to have to do it. You know, they're making profits by killing people. It's as simple as that.'

A picture was emerging of the fossil-fuel producers, be they corporations or nation states that enable them, and then there is the wealthiest 10% of the global population who consume the lion's share of these fossil fuels. It is our spiralling consumption, like a highway into the abyss, that humanity urgently needs to address.

This was only my third COP, but the pattern of failure was writ large. What intrigued me most amongst all this sham politics was that the young people had arrived in Bonn, bringing a new, raw energy. When I spoke to the students Hugh had brought from Cambridge, they had come by car pool and train. Despite three of them not previously having been interested in climate issues, it hadn't crossed their minds to fly.

The students had dived into meetings and been part of the emergent energy. On the last day, I recorded impressions of COP23 BonnFiji from each of them. Through their eyes, it appeared as new to me as it was to them.

Virginia Rollando said, 'It was crazy, I had the chance to speak to the President of the Marshall Islands, who is really worried about climate, but at the same time I met lots of young people who are doing much more than us and are really inspiring. It was intense, and it was great.'

'Generally, I am quite optimistic,' was how Sven Wang viewed things. 'Certainly, big changes need to happen but they are doable. It's really nice to meet a lot of young people and to reinforce that shared value that we really care about these issues.'

Mia Finnamore took the view that: 'I've realised that it is all these people fighting against numbers and figures and politics. I think this is all a distraction from what is really here, which is the planet, the Earth. It's simple. I expected people to be united over that, but it is confusing because people seem to be divided over that.'

The last word goes to Marcel Llavero Pasquina: 'Young people are the future. Young people hold the key to unleash the systemic change that we need. We see it more and more that there is a huge generational gap in how we envision climate solutions. Young people are the ones who can bring this vision that we need – the radical, social, political and economic change.'

I was much more moved by the rising generation than I had expected to be by anything at this COP. It is easy to be cynical, but it is much more important to seize hope when it arises, and here in Bonn, something was rising.

6

COP24, Katowice, Coming of Age(s), 2018

The bus journey from Kraków to Katowice cuts through swathes of agricultural fields and dense woodlands. On the approach to the city, the cold landscape grew gloomier as the gigantic coal-fired Jaworzno Power Station blotted the landscape. This heavy carbon-billowing piece of infrastructure can burn over 2.5 million tonnes of hard coal per year, generating 6.5 terawatt hours of electricity. Katowice has prospered since the mid-eighteenth century when vast reserves of coal were discovered in the area. It felt like an unusual place to be hosting the largest and highest-profile climate action conference in the world.

In his welcome speech to the conference, Polish President Andrzej Sebastian Duda stated that Poland was *not* planning to phase out coal, and that this was not incompatible with climate action. This statement piqued the attention of the world's media. The conference had also accepted sponsorship from large fossil-fuel companies in return for opportunities to promote their businesses, prolonging the use of fossil fuels.

I was reading all this on newsfeeds as I arrived at the bus station just outside the city centre. Disembarking into the freezing cold city, I couldn't help but notice the air pollution tickling my throat and leaving a faint acrid taste in my mouth.

The town itself was pleasant – a mixture of historic and new buildings around the centre, with Christmas decorations making for a brightly lit pedestrianised area. In the taxi to my lodgings, the driver asked me what I thought of the air. I shook my head as he laughed scornfully at the situation.

'Do you think this conference will change anything here?' he asked.

'No,' I replied honestly as he pulled sullenly away from the curb.

Arriving at the COP the next morning, the neon-lit COP24 signage was adorned with delegates, celebrating the moment. The entire venue was abuzz with the usual opening celebrity addresses. Here at COP24, it was Sir David Attenborough from the UK. Sir David, in his ninety-second year, addressed the COP, drawing on his long and broad wealth of knowledge of the natural world. When our planet was smaller in our minds, he was, with his media teams, among the first to expose previously unseen species of plants and other wildlife. Among the latter, many indigenous humans first appeared on our televisual radars thanks to Sir David's adventures. The late British TV and restaurant critic, A.A. Gill, once joked in his column that he had never met anyone who had *not* grown up watching Attenborough on the telly, going on to suggest that Attenborough might well be God.

Here in Katowice, Sir David was embarking on a mission of species intervention, a very different activity to species observation, especially when dealing with *Homo sapiens*.

From the podium, he highlighted a new initiative called 'The People's Seat', which brought people from around the world to share their wishes for a brighter future. It was more grind for the UNFCCC's media theatricals, permeating outwards via the many camera lenses and sound crews but never touching the negotiators in their closeted barracks, tucked away behind us.

The rhetoric was good, such as: 'The United Nations provides a unique platform that can unite the entire world.

As the Paris Agreement proved, together we can make real change happen.' Alas, the reality was more stark. In 2018, the Global Carbon Project calculated that carbon emissions from fossil fuels and industry rose by 2.7%, the highest jump since 2013. CO_2 emissions for that year rose to 37.15 billion tonnes.

Grandiose claims at the start of the COP aim to drive momentum in the negotiations. However, from the outset at COP24, there was no attempt to conceal the motives of the Polish presidency. Their intention was to prolong the burning of fossil fuels, known to be the root cause of the problem, with coal the dirtiest of them all. To underscore the element of tragicomedy, the IPCC's 'Special Report on Global Warming of 1.5°C' was released just weeks before the COP. It stated that the latest scientific research clearly showed that exceeding 1.5°C would be extremely dangerous for all nations. The advice was very clear: *do not heat the planet beyond 1.5°C*.

Instead of the Paris Agreement solving the climate problem, it merely gave us a clear picture of how far we were diverging from reducing the risk of climate chaos. The pathways to stay below 1.5°C meant steep reductions in the burning of coal, oil and gas. In one presentation at the COP, Dr Richard Betts from the UK Met Office stated we are currently on track for around a 3.3°C rise in global mean temperature. This is the range for suffering beyond anything most of us can possibly imagine.

The IPCC's special report on the 1.5°C boundary was receiving political praise from around the world. It was seen as a welcome piece of policy guidance at a time when the global leaders were still prevaricating on taking action.

Sticking two fingers up to the science and everyone pushing for a cleaner energy sector, the United States, Saudi Arabia, Russia and Kuwait actually objected to the phrase of 'welcoming' in the document that was issued by the UNFCCC.

Instead, they used their collective muscle to reword the official response to say they would 'take note' of the report.

This squabbling over specific wording is largely what COPs are about: countries spending many hours watering down texts to remove definite actions, replacing them with more vague *intentions*. Once agreed or conceded, depending on how you view it, the gavel comes down, signing the failures into a toothless agreement that weakens with the memory of the COP itself.

A changing of the guard – 'I don't care about being popular, I care about climate justice!'

In Poland, we witnessed the next phase of a different transition. The one we were meant to be talking about – fossil fuels to renewable clean energy – wasn't happening. However, the slow ebb away from twentieth-century thinking had begun. Sir David Attenborough, despite many of us being such huge fans, was a voice from the previous century.

The young Greta Thunberg appeared in Katowice with Climate Justice Now. She gained notoriety for her 'School Strike for Climate Action' protest that took place every Friday outside the Swedish parliament. The young woman walked onto the celebrity stage with a curious calm. Without histrionics or faux rage, the fifteen-year-old Greta addressed the leaders at the COP, saying:

> You only speak of green eternal economic growth because you are too scared of being unpopular. You only talk about moving forward with the same bad ideas that got us into this mess. You are not mature enough to tell us how it is. Even that burden you leave to us children. I don't care about being popular, I care about climate justice and the living planet. Our civilisation is being sacrificed for the opportunity of a very small number of people to continue to make enormous amounts of money. It is

the suffering of the many that pays for the luxuries of the few. We have not come here to beg world leaders to care. You have ignored us in the past and you will ignore us again. You have run out of excuses and we are running out of time. We have come here to let you know that change is coming, whether or not you like it. The real power belongs to the people. Thank you.

Delivered in a soft-spoken voice, with a few whistles and ripples of applause, it was an unusual speech to be given at a COP. The effervescing energy observed in Bonn was evolving in Katowice.

Scott also hosted Greta and her father, Svante Thunberg, on one of his Climate Show sessions. It was less formal and Scott was in a jovial mood, even making a joke at the outset to which he received a confused look from the young activist.

For me, in this session with Scott and the young Greta, the veneer of so many COP performances began to crack. The cloud of bullshit optimism that hung over these conferences had become a badly soiled joke. No one believed the lies, and as for comedy, only the laughter of incredulity filled the wine-swilling pavilions each evening. The pompous celebrity addresses ceased to stir any emotion. Yet a threat delivered by a teenager skipping school left those with all the power staring back, impotent and exposed.

Scott's boomer generation epitomised the twentieth-century caste that had created the status quo. The male dominance, narcissistic tendencies and bullying impulses, even when deployed for a greater perceived good, are all part of yesterday's world; they are inhibiting progress today. Sadly, despite being a generation younger, I am part of it too, stained by an innate competitive nature that has served us well for thousands of years.

We correlate success with material acquisition fuelled by consumption, accepting a hallucinatory idea of happiness outside the realm of satiety. The need to evolve towards a new paradigm is now required. We glimpsed it that day in Poland. This young woman represented the change that was absent in Paris, searching for its voice in Marrakech, observed in Bonn, and now speaking down to the incumbents at COP24 in Katowice.

There was no abdication of power, as Naomi Klein had referred to in Paris. This was owning the power and directing it back to the people. The press conferences and Climate Shows with the UN-logo backdrops were all part of that old theatre of aligning with the power and trying to move it by tickling its toes. I cannot recall the substance of Scott's session with Greta, but the overall impression of it, as part of a wider mechanism of change, is still fresh in my mind.

Katowice had the grey heaviness of a polluted wintry town, but the people I met in bars, cafés and restaurants contrasted with that. People keenly asked about progress at the COP and pointed to the air, regarding the pollution. It was clear they cared about it. Without a doubt, the air quality would damage the lungs of children, increasing respiratory illnesses. I initially believed that Poland had wanted to host the COP as a signal that they were announcing change. Seventy-two per cent of Poland's electricity power was generated by coal in 2021 and, to the government's shame, the nation has not signed up to the 2050 climate neutrality pledge, as have all other EU states.

With all this in plain sight, was it any surprise when it leaked during the COP that the Polish government wanted to host the conference in order to gain negotiating power to evade any strong commitments away from burning coal? President Duda stated, 'Experts point out that our [coal] supplies run for another 200 years, and it would be hard not to use them.'

These shambolic policy ambitions of the government in no way reflect national polling data. Polish people want clean air and care about the future.

Fear struck cold, Saturday 8 December

I finished working inside the COP at noon. Outside, the air was a cool -7°C, with a grey-white sky softened by sunlight burning through. The path was blocked by hordes of police in heavy body armour. It was congested in both directions. They were like men dressed as cyborgs, hundreds of them. I tried to cut through. The multi-lane street was sealed off. In front were more police in lines. Lots of large guns, some with canisters on their backs, ready to spray. It was an urban siege. I was trying to get closer to take pictures. Heavily armoured vehicles were parked across the roads. Large groups of suspicious men wearing helmets loitered in the distance.

Through the dense armour, a profusion of colours emerged. Non-uniformed people. Civilians with whistles, horns. A drumbeat emerged from the crowd. Police formed blocking lines in front of them. The crowd of protestors, condensed, gaining in confidence, began playing music. They edged forward, establishing a rhythm. The cyborgs edged back. The civilians began a slow march, forcing the boundary.

As I climbed up on the barrier to watch, I noticed more police waiting behind, being filmed, photographed by the press, modelling fear. The cold, the tension, had dissipated from the protestors as they stepped forward, noisily, peacefully. President Duda's brittle ego is displayed, fearful of civilians. The emptiness is the revelation. I walked through the police and into the town, to the Golden Donkey, for it was lunchtime.

'It's nonlinearity, stupid!'

When Dr Richard Betts from the UK Met Office had spoken about the 3.3°C current heating trajectory, he was followed on the panel by Professor Hans Schellnhuber, Germany's lead climate scientist and founder of the Potsdam Institute. He said,

'If you thought this conference can deliver on 2°C, then you have been fooled!'

I had interviewed Schellnhuber in Potsdam just after COP21, back when he was cautiously upbeat and keen that we concentrate on bringing emissions down to the 1.5–2°C global limit. When we met at COP24, I began by asking how he thought humanity had dealt with the crisis over the last twenty years. He said:

> It is actually paradoxical: in the beginning, society was more attentive of climate-change and global warming than it is today. It was a threat where you could play around a little bit. 'Wouldn't it be terrible if it happened?' Today we are in the midst of global warming. You can see it everywhere, and because it is so overwhelming, people just try to push it out of their consciousness.
>
> And this is the problem. We have waited so long to tackle it that we declare defeat, and this is the worst thing that can happen because we cannot solve the problem, but we can minimise it to be something that we can still manage.
>
> If we find reasons to give up, it will turn into an outright catastrophe, and now I know, as a scientist, based on the papers we have published in the last two or three years, that we really face the question whether human civilisation can be sustained over the next century.

I said, 'You have said that we are in a position where we can manage the situation, but on the flip side, we are questioning whether civilisation can be sustained. There is a very stark difference.'

Schellnhuber said, 'Okay, if we get it wrong, do the wrong things – policy, economics, psychology, and in science – then I think there is a very big risk that we will just end our civilisation. The human species will survive somehow but we will destroy almost everything we have built up over the last two thousand years. I am pretty sure.'

I asked what sort of time frame he would put on that.

Oh, it can happen pretty quickly because it is all about nonlinearity. It's the nonlinearity, stupid!

This goes both ways. On the one hand, we can have climate disruptions coming very soon, but in the medium term, if we don't do a lot now, we will send the Greenland ice sheet into irreversible collapse, and so on.

The nonlinearities are our biggest enemy when it comes to the Earth System. On the other hand, why I am still optimistic is that in society, you also have nonlinear dynamics. Tipping points that are social, economic and psychological.

With news of countries undermining the climate agreements at multiple COPs, I asked how people can fight the urge to give up. He said:

Giving up is not an option. Why? Let me give you an example: I have a ten-year-old boy and let's assume he has an accident, and the doctor says, 'Okay, we might save his life if we do this type of surgery but there is only a five per cent chance. Otherwise, he will die!' Would you say, 'No, we don't do it'? Of course, you will do it!

So this is the situation we have now. I think we have more than a five per cent chance of succeeding but it is definitely less than fifty per cent, in my view. But what is the option? If we have a final chance to save our culture and our civilisation, I am just compelled to do it. For the planet, there is no alternative. We definitely have a chance which is above zero.

A chance above zero

The idea of an exponentiating response to the exponential climate and ecological problem was one that I had discussed before with other scientists. Schellnhuber did appear to be thinking along the lines of a range of technological fixes to be developed.

Looking back over the last few years of being at COPs, the techno pathway was often the one that many in developed nations gravitated towards. Perhaps it is because we have benefited so much from technology that we naturally think in these terms. To bring down global emissions to meet the Paris goals depends on human action within a policy framework as the starting point. Without a rapid reduction of emissions, no technology can save us.

In Paris, the saviour was politics and diplomacy. In Marrakech, diplomacy was seen to fail. Figueres adjusted her gaze to business and industry, looking for a momentum shift with Mission2020. A year on from Bonn and the solutions were proving elusive.

In Katowice, the COP was little more than a front for the interests of the fossil-fuel industry. Not just Poland, although they misused the COP presidency to squander the time left to act. In plain sight, countries such as Russia, Saudi Arabia, Kuwait and the United States defied scientific warnings. Other nations, like the UK, Canada and Australia, remained in the background, quietly complicit. The United Nations, undermined by vested national interests, was not working in the interests of preserving the planet for humans or any other life form.

The last three COPs, from Marrakech to Katowice, had witnessed a stirring of anxiousness in youth, but nothing that would qualify as an exponentiating response to meet the exponential challenge.

Scott's news

In April 2019, I heard via several sources that Scott had been diagnosed with cancer and had started treatment. Mike called me, saying Scott was eschewing the medically recommended treatment in favour of a natural alternative. Mike had a conversation with him on the phone and relayed that he was upbeat and confident.

I had not spoken with Scott at this time but I took a moment to reflect on the past few years, his belligerence, and the often maddening way he behaved. We were neither friends nor foes, but his theatrical approach to the COPs was an eccentric element of the annual jamboree. I knew that if he reached his end, I would hear about it somehow. Until further notice, I looked forward with anticipation to see if we would meet at another COP.

7

COP25, Chile [in] Madrid, 2019

In 2019, COP25 was originally planned to be held in Brazil, but after the election of President Jair Bolsonaro, Brazil withdrew, citing economic reasons. The Chilean government stepped up to replace Brazil, but following a rise in metro prices in the capital, Santiago, civil unrest spread across the nation. Secondary-school students refused to pay fares, and tensions escalated. Protestors took over railway stations and subsequently clashed with police.

Since 2015, there had been few signs of progress in aligning with the commitments made in Paris. In 2020, at COP26 Glasgow, nations were scheduled to come together to announce how they were progressing on the road to zero emissions. COP25 was urgently needed to get the climate agenda on track, and with only a month to go, the location was moved to Madrid, Spain. With Chile still holding the presidency, the event was rebranded 'COP25 Chile Madrid'.

Inside the COP, I spoke with a youth activist from the Civil Society for Climate Action in Chile, Angela Valenzuela, to get her perspective. She explained that: 'We've had more than two months of civil unrest. It all started with a protest that is tackling the system of inequality that we have in Chile because it is a system that has failed to provide a dignified life for people and environments, where over sixty per cent of Chileans live with less than minimum wage, and they are paying with credit every month for goods as basic as food.'

The disparity between the politics at home and the presentation of Chile in Madrid amounted to a monumental deception. Chilean officials in Madrid were closer to being stateless than being heads of state. It points to what Professor Dan Bodansky said earlier about politicians making statements and commitments at COP that they might not be able to execute back in their domestic setting. Here was the COP presidency, grandstanding on the world stage, whilst thousands protested 11,000 kilometres away at home.

The huge conference centre was bedecked with stunning imagery of Chilean landscapes and place names in bold type. Commenting on this, Valenzuela remarked:

> For me, it has been very shocking to arrive and see all the greenwashing. All the beautiful pictures, the names of our cities. I knew that COP was a place for greenwashing. That's why it was in Poland last year, one of the European countries that has the most coal plants. But what has been striking is that now human rights violations are happening in my country. It's been overseen by the government and they're really trying to stand out very well and pretend that nothing's happening. I cannot relate or have a proper conversation at the moment with the government. For me, they have lost legitimacy.

Valenzuela epitomised the spirit of action, pushing back against political rhetoric and fantasy. At the time of COP25 in 2019, Chile was already a decade into its worst recorded drought. Scientists forecast it would get worse. I asked Angela how the societal issues she was confronting intersect with the climate issue and how they can even acknowledge the latter, given the current state of unrest. She answered, 'What we are looking for in Fridays for Future and Civil Society for Climate Action in Chile is not that we put environmental issues like the climate crisis as the first priority but at the base of all the demands that are already happening, in terms of minimum wage, pension systems, healthcare and education.'

From an informed youth activist perspective, Angela was articulating how social and climate issues interact with each other in ways that are hard for us to foresee or to control. The corruption of COP24 in Poland by the fossil-fuel lobby, combined with the emergence of Greta Thunberg and youth movement organisations like Fridays for Future, created a space into which broader social issues could find voice.

Touchdown, Madrid

The COP was located one Metro stop from the airport. I stopped to pick up my badge before re-entering the underbelly of the city and riding to the residential Quintana district to check in to my apartment. The Madrid Metro is efficient, clean and well lit, making for easy transit around the town. By the time I went looking for tapas in the centre, it was dark. Entering the vicinity of Puerta del Sol, a city night vibe kicked in. The streets were busy with people in a pre-Christmas mood, pacing bodies cutting between ice-cold gusts of air and wisps of tobacco smoke. A towering Christmas tree beneath an even larger Tío Pepe sign leaned over us all.

On my quest for food, I passed a sherry bar, with aromatic oak casks and unmarked bottles, and an Asturian Sidreria bar, with cider high-poured into glasses held at double arms-length below. Eventually, I found my spot on a stool in a warm tapas bar, pinching the stem of a glass full of bright, ripe, fruity, young Rioja, accompanied by a wedge of the national omelette, a plate of Russian salad, and another of small Padrón peppers, salty and spicy.

As the second glass of crianza touched my lips, my phone buzzed with a diary reminder announcing the World Meteorological Organisation (WMO) briefing taking place early in the morning. Regarded widely as a conservative body, even the WMO had declared the red warning lights flashing on the climate dashboard. I paid up and headed back to my hotel to sleep.

The briefing opened in the morning with the WMO's secretary-general, Petteri Taalas, saying, 'Climate-change is proceeding. We have seen no signs of improvements in the real status of the atmosphere when it comes to the implementation of the Paris Agreement.'

The panel demonstrated with a series of slideshows what was happening to the world to accelerate key climate trends. The trends included record greenhouse-gas concentrations, acceleration of global mean sea-level rise, ocean heat, continued ocean acidification, decline of polar sea ice and the demise of the Greenland ice sheet.

These factors combined are driving high-impact events such as floods, droughts, heatwaves, wildfires and tropical cyclones. In turn, these are creating a deleterious set of outcomes, including risks to human health and food security as well as massive human population displacement, which, from weather events alone, was estimated to spiral to 22 million by the end of 2019. All of this would have been overwhelmingly worrying if I hadn't heard it repeatedly throughout the year and at previous COPs.

Spanish produce

An opportunity to speak with local winemakers from the greater Madrid area about climate-change and their vineyards came in a press release from the Spanish delegation. I called the number at the bottom of the page, set up an immediate appointment and ran like hell to get to the Green Zone where the winemakers were waiting.

I met with Isabel Galindo from the estate Las Moradas de San Martín. Isabel explained how her historic vineyards sit above Madrid at an altitude of 900 metres, benefitting from the freezing cold winters and hot, breezy growing months. They make old-vine Garnacha, a variety that produces fine and elegant wines, easily paired with mouth-watering tapas.

The winters were becoming milder, and the summers much more intense. The rainfall was changing too, becoming less frequent but heavier, crashing down in torrential downpours. The grapes have to be picked earlier, which results in an assortment of complications. These are common signals among wine producers, especially in climates like southern Europe where the intensity of the summers heralds the barren Sahara as it sweeps north, reminding humanity of our folly.

I asked Isabel where I could taste the wines of Las Moradas de San Martín, and she gave me a tip for a tapas bar near the Puerta de Atocha train station called Bodegas Rosell.

Back in the Blue Zone

My phone buzzed with a reminder that Professor James Hansen was participating in a press-conference in the Blue Zone with Scott. Pivoting on my stool, I said goodbye to Isabel, gathered my things and endured the dismembering security processes of re-entering the Blue Zone as quickly as I could.

At the press-conference room, a crowd was forming. Beckwith was waiting outside with someone I recognised as Dr Peter Carter. Peter was based in Canada and published frequent bulletins on the state of the Arctic and other Earth systems.

Within moments, the sound of new arrivals drew our attention towards the entrance of the area. I saw a gang of people heading towards us. Then, in the midst of the crowd, a wheelchair emerged. In it sat Scott. His hair was thinner and grey. At a glance, he might have looked quite weak, but, as they drew close to us, he skipped out of his chair seat and offered me a very cheerful greeting.

'Hey, Nick, it's good to see you. You look well,' he said.

'Thanks, Scott.' I was a little surprised by his spritely emergence from the chair in contrast to his sombre arrival. 'Sorry to hear that you are ill,' I said. 'How are you feeling?'

He leaned in. 'Yeah, you heard about the cancer, right? Look, my doctor said I should be dead already, but I am doing good. You know, I think I am going to beat it.'

We were interrupted by the arrival of Professor Hansen and an additional influx of interested people.

In the press-conference room, I sat in the front row and set up my camera to film. The general content was consistent with Hansen's messaging over previous years regarding emissions reductions and the need to return to a safe atmospheric burden of carbon dioxide. During the session, a woman sat in front of me, almost obstructing my view, even slamming my tripod with her chair at one point. I unleashed some harsh words in her ear and she moved aside a little. The incident added some wobble to my recording.

In a brief interview with Professor Hansen afterwards, we discussed the necessity of researching carbon dioxide removal from the atmosphere. He said that 45% of the nearly 40 billion tonnes of carbon dioxide we are emitting into the atmosphere is being absorbed by the oceans, biosphere, and the soils. He was clear: we needed to get the emissions down and then find safe ways to bring down the atmospheric concentration to beneath 350 parts per million (ppm) of carbon dioxide. In 2023, at the time of writing, we are at 419.39ppm and rising. Although nearly all the historic emissions and much of the current emissions are from developed nations, developing nations are wanting to increase their quality of life, too.

Hansen said:

> Much of the world is still developing, and the undeveloped world has every right to raise their standard of living and they are doing it the same way that we in the West did, by burning fossil fuels. We really have an obligation to work with these developing countries to find the energy that they need without the CO_2 emissions. That is possible, but we are not doing it. India and China would like to have the best technologies, in particular, for nuclear power. They don't want the fifty-year-old technology that we are using in our power plants in the United States. We now know how to burn the nuclear fuel in ways which are much safer, in ways where you cannot have the kinds of accidents you had at Fukushima. So we should work together with them, but we are not doing that. We are even prohibiting the export of technology. That is crazy because we are all in the same boat. We are all going to experience the same fate, so we had better all start working together.

I began by asking, 'There seems to be a big intergenerational divide, where young people don't recognise borders and older people are erecting them everywhere—'

'That is my hope,' he cut in. 'That young people have different attitudes than the older generation. But right now, the politics is still driven by the older generation. the baby boomers, and the young people are saying, "Okay, boomer!" That is a good sign. I think we will start moving soon because of the pressure from young people.'

Thinking back to Paris, when I had spoken with Professor Box, and back to Poland and my conversation with Professor

Schellnhuber, I asked Hansen whether he thought the young people's response would be exponential *enough* to solve the problem.

He drew in a long breath, saying, 'Well, it needs to start soon because the more we put up there, the harder the solution becomes. I hope that in the next year or two we will see a big change. We need to see a change in the United States, that is for sure!'

Afterwards, I went back to the media centre to get the content off the camera and finish up for the day. Working at my desk, I noticed Scott coming in with a reduced entourage due to the need for a press badge to enter. Scott himself never had a press or media badge but was friendly with the staff.

He approached my desk and asked, 'Did you get a recording of that last session?'

'Yeah, not that great, but something,' I said.

'I asked Katie to knock over your tripod to try to stop you filming,' he said matter-of-factly.

'*What?*'

'Yeah, that was me. But, look, it turns out we got a shitty recording. They are using a tiny handheld for cutaways and it is awful. Would you give me your footage?'

I shooed him away, and he settled at a desk nearby with his cohorts, speaking in hushed tones, throwing occasional glances in my direction. Even in decline, Scott was relentless.

Future facing

The UK COP pavilion, having noticed that café culture was making the German stand popular, had offered free coffee at 11.00 a.m. and 3.00 p.m. This was the 3.00 p.m. slot, as I explained when I met Leonie Berners, a young representative from Fridays for Future, the organisation founded by Greta Thunberg.

Fridays for Future had mushroomed since the last COP. Around the world, youths turned out in their millions to strike and march throughout 2019. These fresh activist voices were making the big climate and environmental NGOs look like overly brand-conscious conglomerates. The youth strikers were being obstructive and changing the narrative, stating uncompromising demands to save their future. Walking around the COP, the large NGOs appeared to have joined forces with the status quo. Most had their own big-budget pavilions interspersed among the countries and corporates.

Leonie, an international co-ordinator, gave me an overview of what it was all about:

> Fridays for Future is a movement all over the world. We are striking every Friday for the climate crisis, stating that world leaders have to act, and that they meet the Paris Agreement targets. We strike every Friday to get the pressure from the streets into the government. We are organised country-wise because the climate crisis is a global problem. We cannot solve it just in individual countries. We need to communicate. At the COP right now, we have people from forty-one countries and they all have different expertise. There are also indigenous people that can talk about their suffering from the climate crisis, every day. Then, for example, I'm studying environmental and energy stuff. So I can talk about this.

In the West, we think of 2050 or 2100 when we talk about climate-change, but when you are talking to indigenous people, it has been 'now' and 'yesterday' for quite some time. When I asked Leonie what she thought of this, she agreed, saying, 'I'm really sad when I hear these stories and then I'm back in my country, and people are like, "Oh, yeah, we have still time to solve that climate crisis." But at the same time, I am talking to those that have lost their mother and their brother, or they have no more food, because, for example, hydroelectric dams are flooding their homes, so they cannot fish any more, because they are dying, and all that kind of stuff. Sometimes I feel a bit desperate about this.'

What was evident here in Madrid was that the youth climate movement had caught the media's attention but the process of negotiations was still a closed shop. We were all here but could not see in. Leonie stressed that youths needed to be involved in the decision making as it related to their own lives. She said, 'It's not fair that they decide about our future but we are the ones that have to live with it. I feel panic; I feel it every day when I'm thinking about the climate crisis. I see the climate crisis in every aspect of my life. We need to really take this seriously, to act and not just keep talking about it. We have to really change, and we need an action plan that is controlled every year, that really is a strict climate policy.'

Fridays for Future was founded in 2018, and a little over fifteen months on, the organisation was forming alliances with many social, climate and environmental groups, fighting for issues of justice and change. This young woman looked me directly in the eye and spoke with sincerity. At her age, I had no idea about any of this stuff. Now I was looking at someone who was using every bit of her human agency to end the legacy of suffering brought about by our species, against terrible odds for success.

Dying (for a nap)
I said goodbye to Leonie and hurried across the next giant hall, following the walkway towards the media centre. Turning the last corner, lost in thought, I was caught off guard by a number of people who appeared to be unconscious, strewn about the floor.

I was about to step over the bodies when a voice from someone lying on the floor called out, 'Hey, Nick!'

Clambering to her feet, I saw it was one of my fellow organisers of the Cambridge Climate Lecture Series, Kim van Daalen. Kim was a young medical-health scholar in Cambridge, specialising in the relationship between a changing climate and human health. She delivered the slogan, 'The climate crisis is a health crisis,' and then gave a long list of examples of how extreme climate-change is having an impact on people's health around the world.

It transpired that this kind of action is called a 'die in', similar to a flash mob, except people act as if they are dead to symbolise others who are really dying. I was keen to hear more but my energy levels were running on empty. In fear of reaching an existential crisis of my own, it was time to close this terminally long day at the COP. Recalling the winemaker Isabel's recommendation earlier, I relocated myself to the Bodegas Rosell.

Bodegas Rosell

Exiting the COP centre, I saw a line-up of about one hundred Extinction Rebellion protestors outside, bantering with police officers. With not much going on, I descended to the Madrid Metro and turned on the music on my mobile. The only album I had was Leonard Cohen's last recorded album: a dour-sounding reading of poetry, recorded at the end of a long life, which was a strangely apt soundtrack for this period in Madrid. The Metro was packed with COP delegates, all heading in to the city centre. Many were in groups, noisily excited about being here. My COP foreboding weariness was compounded by the emptiness of my stomach, grumbling at the abuse it had suffered that day, with being given endless cups of green tea and shots of coffee.

The Metro shot down to Gran Via, requiring only one change to the blue line for the remaining four stops to Atocha. Walking down the side of the station, following the phone map, I turned a corner to see the bright red lettering of Bodegas Rosell shining back at me. There is one entrance for restaurant dining and the other, the taberna, for drifters of my ilk, who are greeted warmly by aromas from the kitchen's own Madrid stew. With its stone counter, wood panelling and chalk boards, the taberna could be anywhere in Spain. One board lists regional dishes and the other shows a healthy selection of wines. The floor space is open, with some seating and shelving and an occasional barrel for resting one's glassware and plates.

To quench my thirst, I ordered a small beer, *caña*, which the barman placed in front of me. He then turned away to put some olives and crisps in a bowl and, turning back, removed my empty glass. I added an order of white fleshy anchovies in olive oil and vinegar, and a second plate of cod-in-oil with onions. The white Terras Gauda from Rías Baixas in Galicia, a region I have been visiting for nearly two decades, was the perfect accompaniment. The oily-fleshed fish was cut through by the cool, fresh saline citrus wine, with its fine, long finish. This is the Spain that I love to return to. This simple bar culture fills a tired, solitary person's evening with pleasure.

In due course, I ordered Isabel's Las Moradas de San Martín, Las Luces, from the hills above the city, to accompany pork cutlets in sauce. After I had imbibed and ingested, I had time to reflect on the COP so far, asking myself, what was the point of coming here?

In Poland, there was a meeting of the old and the new. The waning stature of twentieth-century boomers versus a new vision for the twenty-first century. Scott had placed himself at the centre of an intergenerational exchange. Our lot was on the decline. Attenborough stood up in Poland, a voice of the last century, and called on the status quo to change their character. The status quo, Scott, the youth movement, Extinction Rebellion, Fridays for Future, among a growing number of committed people, are all on a collision path exactly because they are refusing to change course.

This new global generation is refusing to transpose those values into this century. The climate crisis is coming for us all because the disruption is global. It doesn't matter who we are. The freak weather, breaking supply chains, the collapse of fossil-fuel infrastructure will impact us all. But how far will we let it go before ordinary people say enough is enough? The thought of it all made me order a second glass of the Las Luces and imbibe some more.

The mood had changed since Paris. The guardians of the system had got too much wrong. The status quo looked weak, tenaciously hanging on. I must hang on, too. Greta Thunberg had not yet arrived at the COP. She had sailed to New York on a private boat for the UN conference earlier in the year and had planned to head to Chile, where the COP was meant to be held. The change of plans had caused some confusion.

Word circulated that the same boat that had carried the young activist to the US was now headed back this way and would land in Lisbon. From there, Greta would make her way to Madrid by train. It seemed like a similar journey to the one I had taken to Porto from Madrid earlier in the year. I took the Eurostar from London to Paris and then another train to Barcelona. After a night there, I travelled to Madrid, dined and then took the sleeper to Lisbon, changing in some small, nondescript town for a regional train to Porto. Travelling in the old way is fun and adventurous. A new high-tech rail network in Europe could easily replace the undignified discomfort of the short-haul air industry.

I was warning myself to stay vigilant and pursue these emerging signals of changes that were happening, not at any one COP but across all of them, in succession. To achieve this in any meaningful sense, I would have to make this my last glass of wine.

Rock'n'roll star

The next day at 11.00 a.m., I was sipping complimentary coffee on the UK pavilion when Kim from the 'die in' walked by. While chatting, she said that she had heard that Greta Thunberg was arriving at the COP. The audible volume of activity was growing. Activists and indigenous peoples were moving past us and assembling some way off. As we were talking, I checked my phone emails and saw the following from the UNFCCC media centre:

> Media Alert 2 – Greta Thunberg is currently at the COP venue in Hall 4. She may be speaking in the Amphitheatre shortly. Whatever she does in the Amphitheatre will be pooled. This is advice I have just received. Not confirmed.
>
> Regards,
> UNFCCC IBC Manager, COP25, Madrid

As I read this out to Kim, we looked up to see a herd of cameramen and press people charging like wildebeests past us, across the gangway, into Hall 4.

It was as if Michael Jackson or The Beatles, at the peak of their fame, had entered the building. A tight media scrummage with teenagers crammed in the middle formed a high-pressured core. UN COP police officers were trying to hold people back, but it was futile. The scrummage started to move; the youngsters were still jammed inside and, I expect, trying to breathe.

There was something savage about the scene, the way the hungry pressmen sought their prey. On the flip side, there was something fascinating about how we create our heroes by smothering them in adulation. There was something of the Messiah about this moment. As more UN police arrived to break up the crush, the young activists were whisked off to a safe hideaway.

Sharing the platform

Fridays for Future compounded the frenzy by announcing a press-conference for later that day. I turned up to watch, but the mob was already there, forming a thick line through the hall. Avoiding the crush, I ran back to the media centre and watched the whole presser on the UN livestream.

The Fridays for Future team appeared spread out along the stage with German youth activist Luisa Neubauer and Greta Thunberg at the centre. They acknowledged the interest they were attracting and instantly handed the microphone to representatives of indigenous peoples present on the panel.

Carlos Zackhras from the Marshall Islands was the first to give his story:

Before I came to Madrid, exactly two weeks ago, I experienced sixteen foot [4.87 metres] swells that forced 200 people from their homes. Not only do we have inundations but we also have the dengue fever epidemic, the flu, and I know just recently, our Pacific neighbours in Samoa are fighting the measles, which took seventy lives, thirty of which are children under the age of four. These are illnesses linked to and made worse by climate-change. We have been told that if we want to stay on our island, we would have to adapt and elevate, with migration as the only Plan B. And may I remind you, the Marshall Islands' contribution to climate-change is only 0.00001% of the world's emissions.

The next speaker was Angela Valenzuela from Chile, whom I quoted at the start of this chapter. Valenzuela addressed the packed room, saying:

> The cameras of the world don't often look at the Global South. In the case of Chile, since COP25 was suspended in Santiago, most mainstream media forgot about us. Millions of people marching on the streets for a dignified life. It's not about thirty pesos. It's about thirty years of democratic governments that failed to protect us and to listen to our demands. In order to address the climate crisis, human rights must be protected. While countries congratulate each other for their weak commitments, the world is literally burning. Over the last week of negotiations at COP, instead of talking about how to transition from fossil fuels, COP25 is focused on finding elaborate ways for rich industrialised countries to carry on polluting, while pretending not to. People are already dying in the climate emergency and these communities need support. But again, it seems that some lives matter more than others. The rich and powerful seem happy to sacrifice our communities in the pursuit of profit. So, here at COP, back in Chile, and around the world, people will continue to rise against governments that do not represent us. Our lives are not up for negotiation. Our planet is not for sale.

The stories continued from other young people from far-flung places, giving voice to new faces, cultures and plights. It was not about the heroism of one person but the coalition of the many to call for system change. Not that this would happen here today on stage, but it was more a signal that this was the start of the change.

Valenzuela had been impressive in the press-conference, declaring 'we are fearless' and, with regard to the ongoing civil unrest demanding an end to the oppression, 'we are redefining our future and pushing the limits of what we think is possible!'

The richness of these peoples makes up the throng of the crowds here at the COP, but seldom makes it into the realm of the narratives emanating from it. I could not escape the sense of foreboding that my people, in the West, were responsible for the suffering, the theft and the injustice that lay beneath the surface of greenwash. The climate emergency was the sum of our actions and, listening to these stories, it was clear that the Global South is our victim even to this day. We like to say we are good people and that we care, but when it comes to cutting back on consuming, whatever the whim, we look the other way. When we see suffering on the news cycle, more likely on social media because the corporate news omits to cover it, we make the right noises and carry on as usual.

Only now, the scientists are more worried than before because extremes we are experiencing indicate the Earth is more sensitive to carbon pollution than previously thought. The impacts we are experiencing at lower temperatures mean the Global North is going to suffer too. Not just the next generation, but ours, too. Forget the twilight years of old age or growing old gracefully. The era of shocks and high-risk impacts, from food shortages and energy blackouts to cascading extreme weather events, is coming.

Madrid Climate March
A tall, bespectacled figure flagged me down as I trotted down the concourse. I recognised him as Portuguese Carlos, a climate activist from Lisbon. We met previously in Porto, where he was working with Climate Reality. He was assisting Scott here in Madrid. 'Are you going to the march?' he asked.

'Yes, leaving now. You?'

'Mind if I join you? There's a group from Climate Reality meeting in front of Atocha. Perhaps we can catch up with them?'

'Sure, we can try, but it is going to be very busy.'

The Madrid Metro, from COP to the Estación del Arte, at the front of the Atocha train station, was crammed with people – a human press. The wise peeled off along the route to escape the squeeze. Emerging into the street, the noise was deafening. Music, chants, drumbeats. My palate was dry. It was Friday and a small tapas bar, La Gloria, lurked in the corner. We entered the bar and ordered two small beers and two Spanish omelettes, tortillas. We ate at speed. As Carlos tried to locate his Climate Reality group, I watched through the bar window, enthralled by the endless stream of people passing. It was a steadily building tide, and now was the time to dive in and swim with it.

The protest route ran along the Paseo del Prado, next to Spain's national Prado Museum. The energy was full of carnival. We moved at speed to get ahead, to try to catch the speeches and to meet the others. The crowds stopped and started. Protesters had made huge efforts, coming from across Spain, across Europe and from the farthest corners of the world, dressed up elaborately, to be seen. At the front, indigenous peoples of Chile led the march. This was a signal, acknowledging the human rights abuses being reported, including torture and killings.

We hiked up a narrow lane, left from the Neptune Fountain, just past the Prado, and tucked into Los Gatos, a small Basque-style pintxos bar, thick with aromas of oil, fish and white wine. A scene from the Sistine Chapel is painted on the ceiling; it is a cosy place, Baroque in style, busy with people ordering fishy pintxos from a glass bar-mounted cabinet. We relished the anchovy fillets on thin bread, washed down with more *cañas*.

When we re-entered the fray, the crowd was even larger, but the sun had gone down, leaving only darkness and the pounding beats as our guide. The masses appeared as shadows with smatterings of pale flesh, punctuated by sombre red rebels. The uniformed watchmen kept their vigil at the edges, grouped at the main junctions.

Approaching a clearing where police were in particularly large numbers, a column of young protestors, all in black, ghostly shadows, filed between the dancers and musicians, attacking the police with missiles. The crowd scattered. I tried to film what was happening but people were running all around us. It felt staged. Carlos disappeared, ran with the crowd. I spoke to a man with a cello and he scoffed, 'Niños!' The march proceeded.

At the finish, the crowds were intensely focused on the stage. Greta appeared as a small dot in the centre, rousing her people. Javier Bardem, singers and dancers, emboldened protestors. Was this a movement galvanised? Or simply a nation that loves to party? ¡Viva España!

We, the people
Thunberg addressed a 'high-level' meeting at the COP, appearing at the podium flanked by the flags of the United Nations, of Chile, and of Spain. If her expression hinted she would rather be somewhere else, her words demonstrated her awareness of the futility of speaking here:

> Hi. A year and a half ago, I didn't speak to anyone unless I really had to. But then I found a reason to speak. Since then, I've given many speeches and learned that when you talk in public, start with something personal or emotional to get everyone's attention. Say things like 'our house is on fire', 'I want you to panic' or 'how dare you?' But today, I will not

do that because then those phrases are all that people focus on. They don't remember the facts.

She followed this by quoting the compounding climate issues that have echoed across these pages from the mouths of numerous scientific experts. Then, instead of pleading with the politicians, diplomats and negotiators, Thunberg bluntly pointed out the failure we could all see:

> All leaders are not behaving as if we were in an emergency. In an emergency, you change your behaviour.
>
> If there's a child standing in the middle of the road and cars are coming at full speed, you don't look away because it's too uncomfortable, you immediately run out and rescue that child.

The final bit of the speech addressed the application of agency. It was a plea, not to the leaders, but to global citizens to seize back the agenda. The youth movement was spreading, and this young woman was setting an agenda to reach beyond the horizon of COP26 Glasgow.

> Right now, we are desperate for any sign of hope. Well, I'm telling you there is hope. I have seen it, but it does not come from the governments or corporations. It comes from the people. The people who have been unaware but are now starting to wake up. And once we become aware, we change. People can change. People are ready for change. And that is the hope, because we have democracy and democracy is happening all the time. Not just on election day, but every second and every hour. It is public opinion that runs the free world. In fact, every great change throughout history has come from the people. We do not have to wait. We can start the change right now. We, the people.

The media focus was shifting away from the UN and the wider political rhetoric. Naomi Klein's Christiana Figueres quote – 'Never has so much been in the hands of so few' – appeared to be inverting before our eyes. The bureaucrats' handle on power was slipping as a new consciousness was emerging. 2019 felt very different to 2015. The masses outside in high-polluting nations weren't all getting it yet. There was not enough urgency to interrupt a dinner party or cancel a holiday to a favourite resort, but anxiety was building in the collective subconscious. Survey data showed that the majority of Britons regarded climate-change as their biggest concern. In the US, numbers were also rising. Young people grasped the immediacy and scale of the challenge and were connecting across borders. The incumbent baby boomers remained content to keep on burning the house down. A few years down the line, those same youngsters will look us in the eye with disdain.

Final hours

The next day was my last in Madrid and I had planned to interview Dr Peter Carter and a winemaking legend, Miguel A. Torres, head of one of Spain's largest and best-known wine-producing companies, Familia Torres, and a climate-action evangelist.

I had interviewed Señor Torres previously at the wine and climate event in Porto and had also attended a small conference on the same subject at their headquarters in Catalunya. A commitment had been made throughout the business to try to achieve true sustainability and implement climate-safe policies. No one company in any sector can do it alone, so Torres joined forces with Jackson Family Wines in the USA, starting International Wineries for Climate Action (IWCA). This collective of wine producers are measuring their carbon impact and sharing knowledge across the IWCA network to build a resilient future-facing industry.

The size of the Torres business and their profile on climate action meant that it was no surprise they were represented at a COP on Spanish soil. The wines and brandies are internationally famous and emblematic of Spanish culture.

Señor Torres described the pressure the vines are under from the heating climate. The impacts on vine health and grape quality are now all too familiar. When I asked about his level of optimism, that people will change, Torres answered, 'The vineyards are like a thermometer for climate-change or climate emergency. We see the impacts constantly, but for most people it is still not a priority. Most people, they just continue with business as usual. They continue to fly, to drive their cars, to eat meat all the time. So, very little changes.'

I said goodbye and wandered off, contemplating a world where survival was threadbare and wine was a distant memory. A place of great mental and physical discomfort.

The intersection of climate and conflict

Running through the pavilions on my way back to the media centre, I received a message from a young environmental lawyer I had been introduced to a few days previously, Shirleen Chin. Shirleen is the founding director of Green Transparency, a boutique environmental law agency based in The Hague. She is also closely involved with the Ecocide campaign, which has gained international recognition for its efforts to make environmental damage a crime against peace at the International Court of Justice.

Shirleen was doing some work with a group at the COP called the Global Military Advisory Council on Climate Change (GMACCC). GMACCC consists of a group of international military and security experts, including former ministers, retired high-ranking military officials, and academics, with a mission defined on its website as 'an authoritative voice on the need for, and the nature of, action to be taken by the security community to reduce the risks posed to national security by a changing climate'.

For a petite lady, Shirleen is certainly a larger-than-life character, vivaciously directing everyone in her orbit towards the next relevant conversation. She flagged me down and ushered me into the GMACCC session in progress, titled 'Climate and Security – Emerging Trends and Adaptive Strategies', specifically focused on the challenges being faced by Least Developed Countries (LDCs) and Small Island Developing States (SIDS). The discussion dealt with some of the issues surrounding preparedness for changes that were hitting countries where communities were both poor and ill-equipped to respond to extreme climate impacts.

After the session, Shirleen introduced me to one of the panellists, former Pakistani Defence Secretary, Lieutenant General Tariq Waseem Ghazi. The role of military organisations in climate issues is a subject that is discussed regularly. For a number of years in the US, the military has cited risks associated with changing climate as a high-level threat multiplier in relation to national security. This is despite the US administration of 2019 espousing a childlike denial of this obvious and ever-growing reality.

I asked General Ghazi for an impromptu interview to run over a few of the points he touched on in the panel discussion and began by asking him about the role of the military in places like Pakistan.

General Ghazi stated that:

In disasters, it is the military that become the first responders. I know that in Pakistan, and I think in the rest of South Asia as well, the military is trained to prepare for disasters. We know that in the monsoons, there is going to be flooding in certain areas. The trouble is that climate-change is now making it very erratic. The monsoon may not come in June, it comes in September. Suddenly you have to deploy yourself when you might already be deployed in something else. So there is this resilience that is needed in the manner in which we can assist people in disaster management, and in rehabilitation.

The destructive nature of climate-driven extremes came true for Pakistan in August and September 2022 when they endured a sixty-day period of intense rains. Horrific images of the most catastrophic flooding in Pakistan's collective living memory started appearing on news outlets around the world. The floods decimated communities: 1,700 people were killed; 33 million lost their homes; the economy was left in ruins. An assessment a year later found that the country had not been able to recover. Haroon Janjura, reporting in *The Guardian* from Islamabad in August 2023, stated, 'Forty per cent of the children they surveyed had stunted growth and twenty-five per cent were underweight as families struggle to access food and healthcare. About eighty per cent of mothers reported sickness among children, with outbreaks of diarrhoea, malaria and dengue fever increasing.'

The sixty-day period of intense rains was turbocharged by increased carbon pollution, of which Pakistan is responsible for just 0.8% of the global contribution. The atmospheric burden of greenhouse gases is still being made worse by countries and blocs like the UK, USA, Canada, Norway, Australia, China and the EU, among many others, delaying the much-needed rapid phasing out of carbon-polluting energy.

Here in Madrid, 2019, I asked General Ghazi what challenges concerned him the most at the intersection between climate disaster and military responses. He answered:

I think my biggest concern, especially in the areas in which I live, is water. It is the reduction in the flows of the River Indus on which people in my country depend, through either human activities that are diverting waters and making the lower riparian suffer. Now, if the intention is to make the lower riparian suffer, because there are certain geopolitical or political reasons for which you want subservience or compliance, then I think it is the trigger for a catastrophic conflict. Because in the end, when you have no security of livelihood left, then the only option is to resort to arms to actually retake it. And that is something that I see happening in terms of water threat.

In this direct reference to the growing tension between India and Pakistan over the damming of the River Indus, with a potential to severely restrict vital water supply to communities lower down along the riverbanks, the pathway to military confrontation becomes much clearer. Quietly, I noted to myself that instability between two heavily armed nuclear powers could very well be the trigger for catastrophic conflict and much more suffering.

Dodgy carbon offsets and double accounting

As the COP entered its final days, rumours of setbacks and delays were circulating. I arranged to speak with an analyst from the World Resources Institute, who concluded this COP would be a 'technical failure', as key points of the agreement sought were being kicked down the road to next year at COP26. As I listened to the cause and effect, it seemed more removed to me than if the negotiators had been trying to decide whether to bury or cremate victims of climate impacts. The reasons for the stalling of agreement were preceded by statements along the lines of 'all parties are calling for ambitious climate protection targets' in order to 'enhance the ambition to reduce carbon dioxide emissions in countries all over the world'.

Totally bizarre. The 'technical failure' was attributed to the fact that negotiators could not agree on how carbon markets should operate fairly without double accounting emissions. The latter bit is funny-speak for 'bare-faced lying about emissions'. Double accounting is not an accident. It is a design feature of countries who like to mislead us on the world stage by still claiming to be making progress.

When I spoke to Dr Joe Romm, a Senior Research Fellow at the University of Pennsylvania's Penn Center for Science, Sustainability and the Media, he gave this overview of emissions double accounting:

The double accounting problem is quite simple. Imagine we have the United States and Brazil, and I'll just use a very simple example, where, let's say, Brazil had two billion tonnes of CO_2 emissions a year, and the United States also has two billion tonnes. So together, we have four billion tonnes. Now, the United States says we're going to pay Brazil to reduce its easiest emissions reductions. So we are going to pay them to cut a billion tonnes of emissions by tree planting, shutting down some coal plants, that sort of thing.

So Brazil takes the US money; they reduce their emissions from two billion down to one billion. Now, the total emissions in the world are three billion. If, however, the United States said, 'We paid for those emissions reductions, so we are offsetting a billion of our two billion and we are going down to one.' All of a sudden, the total is two, right? The US has a pretend billion and Brazil has a real billion. So we are claiming that between us we have two billion, but only one billion was actually really reduced. They have just doubled the amount of emissions reductions. So that is double accounting.

COP25's double-accounting failure was revisited at COP26 in Glasgow and again the following year in Egypt. Joe summarised the outcome of these developments, saying:

> In COP26 in Glasgow, the world agreed to two things. The Paris accord was going to set up something called an authorised offset. This authorised offset was going to be the real deal. The purchaser of this offset gets to say that their own emissions are lower. So that requires the seller to publicly announce to the world officially on record: 'I'm not counting these emissions reductions, I'm going to add them back.' And the adding back is something called the 'corresponding adjustment'.

One failing of the system that Joe points out is that, under the Paris Agreement, countries and large corporations are not restricted from mixing up carbon offsets that are bought and sold at a fair price with no double accounting and those that are sold on the dodgy, unregulated 'voluntary' carbon markets.

Joe highlighted that this arrangement exposed the poorer countries to a form of climate colonialism because wealthier nations can tempt nations who need funds to sell their carbon emissions reductions at a far lower price per tonne than their real value. As we move forward into an era of real emissions reduction and proper accounting, the price of carbon per tonne will rise as it gets harder to keep finding emissions to reduce. If all nations have committed to getting to zero, a poorer country selling the authorised offset and making the corresponding adjustment now, will eventually have to find those emissions reductions from somewhere else. The only place left will be an official carbon market, called a 'compliance market' (because participants are complying with international regulations).

His research on carbon offsets led to the conclusion that they are nearly all dodgy and, ultimately, developing nations might have little choice (and be justified) in annulling the cheap tonne deals they made earlier on. Deals are already being brokered now for cheap tonnes on the voluntary markets by large corporations and nation states that are making a mockery of the Paris Agreement. In one example, Denmark's government paid for a Danish fossil-fuel company to clean up its emissions. It then took the offset value of those emissions and added them to its national emissions account on the compliance market. The oil company (majority owned by the state of Denmark) then sold the same emissions again to Microsoft on a voluntary (unregulated) market. This was not illegal but the morality is certainly questionable.

This practice of mixing carbon offsets on the compliance market with the voluntary market could have been resolved at COP27 in Egypt, when the negotiators looked at this specific deal but did nothing. Joe finished, saying, 'No one can tell the voluntary market what to do. But the Paris Agreement people can tell the nations who are signed up to it what they can't do, but they kicked it down the road. The issue was: does a voluntary offset need to have a corresponding adjustment attached to it? And they said, "We are not going to decide." But the correct answer is: they needed to decide! Every single nation needs to step up and stop this gaming of the system.'

In moral terms, it's evil

The clock was ticking as I walked briskly back to my desk in the media centre. This being my last day in Madrid, I had to try to upload what footage I could before packing up and making the short sprint to the airport for my estimated personal 124-kilograms-of-carbon flight distance to London. Nearing the group of desks where I was working, I rounded a corner and nearly sent Dr Peter Carter flying.

'Ah, Nick,' he said. 'Good to see you. Are you still up for recording the interview?'

Of course I was, despite having been overwhelmed by the ticking clock and other relentless checklist items. I invited Peter back to my workstation, where I wrote a structure of questions for him to answer, which would also help to clarify my own thoughts at that moment in time.

Before meeting me, Peter Carter had been at an enclosure called the 'Action Stage' where Greta Thunberg had introduced a small group of climate scientists. Her stated reason was that there was no science in the COP negotiations. Peter was explaining it to me:

So last night, I went to the Climate Secretariat and checked all the documents that they have up to this day, and there is no mention of science. No mention of the IPCC. There is no mention of the IPCC 1.5°C report. No mention of the two reports that were mentioned by the scientists that I just heard, referencing the most important reports on the cryosphere (the frozen regions of the planet) and on the oceans. Then no mention of the other report about climate-change on land, which is most important because it went into impacts of global climate disruption on crops, namely food security. So this is very, very bad.

Last year at COP24, there was a lot of media attention given to the terrible fact that four of the countries got together to block the tabling of the most important IPCC report ever, which was the 2018 IPCC report on 1.5°C, which showed that the old target, since 1996, is total catastrophe, and 1.5°C is still disastrous, but that is where we must aim. All the scientists now are agreed on that. They're all behind 1.5°C, although it seems almost impossible to do that, but the scientists today said no, it still is possible, only if we reduce emissions by seven per cent every year from next year.

Peter diligently outlined our current position as being the climate-change worst-case scenario. We needed to reduce emissions by 7% per annum from 2020 to achieve a 50% reduction of total emissions by 2030.

In 2012, when I first interviewed Professor Jim Hansen, he said that if we didn't start reducing our emissions in 2012/13 then trying to start in ten years' time (2022) would be very difficult because that would mean 10% per year. Yet the global polluters have done nothing. His vocal outrage in Paris in 2015 didn't reach a single policymaker. The COP process was like a giant condom, trapping all the toxins and fertile information inside, preventing the truth from making contact with the outside world. The general public had no idea their lives and loved ones were being cast into the furnace for economic growth and profits, for the short-term benefit of a few faceless people.

When I asked Peter whether he thought the COP could solve the crisis we are in, he said:

It has always been set up to fail since 1995. The first two COPs were pretty hopeful. Ever since then, things have gone down, down, down. The reason why they are set up to fail is it said that major decisions will be made by consensus. So the first thing in 1993 that these negotiators, the parties, got together on is, what are we going to call consensus?

We still don't have a definition of consensus under the convention. Every single year they have what they call an 'ad hoc decision' and the decision is made by whoever is hosting and holding the COP meeting, like how the French government made the Paris Agreement. They said that there is going to be total agreement or there is going to be no agreement, or virtual unanimity. In other words, they'll pass something if one or two parties object.

Now we know for sure that the United States, Russia, Kuwait and Saudi Arabia, at the very least, are blocking the science from the negotiations, and it seems that in this COP the science has been dropped completely.

The next question was obvious. What would the negotiations sound like if the science was injected into them? Peter replied that we would drop the language of 1.5°C or 2°C and use atmospheric greenhouse-gas concentrations. This was also what Dr James Hansen had said back in Paris. He then pointed out that the greenhouse gases – carbon dioxide, methane and nitrous oxide – were all increasing in both their concentrations in the atmosphere but also in their annual quantity of emissions. I asked him what he meant by that, and he explained:

It means we're on the trend to total planetary catastrophe; we are on a trend to biosphere collapse. And this year, 2019, the atmospheric carbon dioxide increase rate was the highest that it has ever been. So we are increasing CO_2 in the atmosphere at a rate the World Meteorological Organisation has told us is definitely faster than anything that has happened in the past 40 to 50 million years.

Peter then took me on our familiar tour of Al Gore's 'nature hike through the Book of Revelation'. He finished by saying:

What is happening to mitigate this? What is happening to respond to this? What is happening to prevent all of these forest fires, droughts, severe storms, powerful hurricanes, floods? What is happening to lessen them at least? Absolutely nothing.

Nothing is going to come out of this COP. On the very first day, the parties made the decision that they're not going to look at an improvement of their national emissions targets. Now, what do you call this? This is a terrible, terrible crime.

It is unbelievable what these high-emitting fossil-fuel-producing countries are doing. Pope Francis said it is a sin against God and, very recently, he said it is a crime. That means, in moral terms, it is evil. So the result of this is evil. The countries that are blocking any progress on emissions are acting in the most evil way that anybody could imagine, because we are looking at the destruction of Earth, our oceans and land.

This was a dour end to the Madrid COP. It was already breaking in the mainstream news that it was being labelled a failure. The carbon market non-issue was being kicked down the road to the mega-COP26 Glasgow, where all nations planned to get together and announce progress on emissions reduction based on their NDCs (Nationally Determined Contributions).

Back in Paris, the Potsdam Institute calculated that the NDCs added up to climate catastrophe. In the previous four years, we appeared to have been going backwards on climate action and emitting more than ever before. What was a catastrophe then has been getting much worse.

I gathered my things and began the long walk towards the exit from the media centre. It was contemplative. You can't come to a conference like this, listen to the facts, and leave without a nagging despair lodged in one's mind. I made my way through the corridors with the weight of my bags causing me to break out in a sweat.

Half of me wanted to dump the stuff and wine the night away in deep thought in a Madrid bar. The wiser half knew that a happier ending was in getting back home and crawling into bed with Natalia, no matter how late, and falling asleep.

In this surreal journey across the halls of the COP, I heard a familiar voice call my name. 'Hey Nick, are you off now?'

I turned to see Scott sitting in his wheelchair, hands on the armrests, looking relaxed.

'Yep, I'm done here and heading back to London.'

The show was over. He hopped, with agility, out of the wheelchair and shook my hand.

'Travel safely, man. Good to see you.' He was cheerful. 'Oh, can you *please* send me that footage?' he added with an instant pleading face and a half smile. I was surprised that he actually used the word 'please'. I said I would send it when I got home.

We said goodbye.

8

Post–COP Global Disruption, 2020

The Madrid COP had failed to complete its business. The hype for Glasgow began early in 2020, amid rumours of the UK presidency being shambolic. COP26 Glasgow was being billed as the next megaCOP after COP21 in Paris. The expectation was for nations to show clear progress on their NDCs to reduce carbon emissions before ratcheting up their ambition in order to bring their pledges in line with the 1.5°C boundary.

The clock was ticking. Hundreds of billions of tonnes of carbon emissions had been released into the atmosphere since Paris. The window for achieving what scientists refer to as 'a safe operating space for humanity' was rapidly closing.

In early 2020, the world was gripped by the COVID-19 pandemic. At some point during the lockdown, I saw a photograph of the Madrid conference centre where the COP was held, converted to a vast hospital for people with the virus. My memories of the corporate exhibition spaces, the wildebeest press, die-ins and other COP antics clashed with this dystopian image.

Many people engaged in climate issues anticipated that a single benefit of the global lockdown would be reduced carbon emissions, as the world ground to a halt. Actual global carbon emissions fell by an estimated 6.4% in 2020, compared to 2019, due to the pandemic, according to the journal *Nature*. Unfortunately, emissions from fossil fuels bounced back by

5.3% in 2021. In 2022, global greenhouse-gas emissions from fossil fuels and industry rose again by 1.5% to 36.6 billion tonnes, as reported by *Nature*.

As the death count from the pandemic increased, it became obvious that the much-anticipated COP26 in Glasgow would be postponed. It was rescheduled for November 2021.

In July 2021, I received news via several sources that Scott had succumbed to cancer, dying at his home in the US. I reflected on the seven years I had known this eccentric character. I doubt there are many people who attended the huge number of climate events that Scott frequented who could forget his intense yet charismatic demeanour. He was aware of his over-the-top-ness and often made fun of himself. We certainly fell out on several occasions but none of the malice stuck. Each time, anger was quickly replaced by head-shaking disbelief. In one of his last emails, he stressed the toll that his unorthodox combination of hustle and activism had taken on him. He appeared to thrive on pressure, even at the end, when his battle for climate action was expanded to include prolonging his own life.

Would the COPs ever be the same, or would the COPs' now deceased Climate Show host manage an intervention from beyond the grave? I doubted it, but it was still hard to imagine what a COP would be like without bumping into Scott.

9

COP26, Glasgow, 2021

My first impression of Glasgow was a collage of dark Victorian grandeur and worn stone edifices, dissolving into bright, wintry shafts of sun, illuminating posters, installations, advertisements for concerts, and environmental slogans. The COP bus rolled through George Square, the hub of the city, a bright collage of protest signs, colourful faces and indigenous costume. Everywhere the bus turned, there were people in abundance. My ambivalence about coming to the COP contrasted with a feeling of excitement for the spirit of Glasgow.

I had arranged to spend the evening with an artist friend, Alex, who promised to show me a typical Glaswegian pub. Before heading out, Alex produced a box of Bordeaux wine samples that had been sent for me to taste. We opened a bottle of the delicious Château Guiraud Blanc to enjoy with a fresh shellfish chowder Alex had prepared. My level of inebriation was nothing compared to the general standard set by fellow patrons of Alex's local pub. His friends had been there all day and were showing signs of strain. One in particular had a routine of pulling me aside every ten minutes and staring disconcertedly into my eyes before bellowing, 'I've beeeeeeeen to the Ecuadooooorian Embassssy! I'vvvvvve met Julian Assange!' I nodded enthusiastically at this declaration, each time as if it were the first. His Brazilian friend stood next to him and glared at me with his own sombre intensity. He, too, it appeared, understood the significance of the declaration. We all exchanged deep empty stares before the spell was broken by the arrival of more drinks.

In the morning, Alex dropped me at my hotel, a glass-fronted construction built into the side of the Glasgow Central train station. This was to be my new base for a fortnight.

After a quick check-in, I decided to walk to the COP, which was along a straight road on the banks of the River Clyde. It was an overcast day but not cold and, when the sun burst through, the shapes of the clouds remained stencilled on the backs of my squinting eyelids. Glasgow was once known as the 'second city of the empire' due to the volume of its trade with America. Today, the Clyde swept past me as I walked beside low-rise residential blocks and a tempered modern urbanity.

The banks of the river brought me to the outer perimeter of the COP. The police presence felt like an army of occupation, with every officer kitted out for conflict. By contrast, the humans beneath the armour were very friendly and helpful. Confusion reigned as delegates realised they had to be tested for COVID and be able to show the test result on a government-branded mobile application. The queues to get this sorted out were snaking down the road and I fretted about waiting for hours in potentially COVID-infected company.

I remembered I had some testing kits in my bag. Taking one out, I sat on a bollard in the street to test myself. A young woman in a high-visibility jacket came over, a local, drafted in to help oddballs like me attempting to do odd things in the street. Assuming control, she had the whole test completed in a few minutes, whilst also telling me everything about her life, including a number of siblings, previous jobs, how many idiots like me she had helped in the last few hours, and when her shift was ending. During my stay, it transpired this is a Glaswegian trait. Hospitality in the city was plentiful and always framed with disarming chatter that could go on indefinitely. I gathered my pass and returned to my hotel. An evening of fish-and-chips beckoned.

Day 1: Bedlam and bad jokes

I arrived at COP at 8.00 a.m. and was stuck in a human traffic jam, each person as anxious as the next, hoping there was not a new strain of COVID in the making.

Finally, I got through and made the long run towards the media centre. COP planners must despise the media. Every year we are located at the farthest distance that is possible, meaning it takes an extra twenty minutes to walk to the designated area, even at very high speed, weighed down with equipment.

I had booked a working booth for *The Ecologist*, who had a group of reporters working here inside the COP. The booth could be locked with equipment and left overnight, which made life a little easier.

I detected the mixed aromas of enclosed bodies, all hovering around by the entrance to the main conference hall. The press frenzy was building, with cameras, reporters, general COP tourists, all waiting to glimpse the A-listers as they entered for the opening speeches.

This was the familiar theatre of the COP, in which grandees and carefully selected VIPs get to make statements, imploring the conference to work its magic (as yet never achieved). What was strange about the crowd gathered outside was that it was so energised, yet, at best, we were likely to glimpse a bunch of faceless bureaucrats and a few minor celebrities. The really big names came in through the back door. I returned to the booth to watch the live feeds.

On the screen in the corner of the booth, a dishevelled Prime Minister Johnson stood before the world's media. Given the gravity of the situation, he was the perfect person to confuse the conference with his brand of rhetorical blather. And so he did:

> Welcome to COP, welcome to Glasgow, whose most globally famous fictional son is most certainly a man called James Bond, who generally comes to the climax of his highly lucrative films strapped to a doomsday device desperately trying to work out which coloured wire to pull to turn it off, while a red digital clock ticks down remorselessly to a detonation that will end human life as we know it. And we are in the same position, my fellow global leaders, as James Bond.

There was another eleven minutes of this speech, but the standard of gibberish had already been well set. After challenging the negotiators to 'not sit on their hands', he returned to his seat and was recorded folding his arms and nodding off to sleep. Johnson then demonstrated his disdain for efforts to reduce carbon emissions by flying, in a private jet, to London to dine at a private members' club with a reputed climate-change denier.

Sir David Attenborough, an inevitable British voice for the natural world, graced the stage. I couldn't help thinking that here we were, a million miles from any meaningful action, watching a parade of dodgy politicians and showbiz entertainers, leading on a tired act. Attenborough's stern speech was a warning that ended on a hopeful note, extolling the ingenuity of our species. I thought to myself, if this ingenuity exists, then why so coy?

One speech set the facts before the crowd. Barbadian Prime Minister Mia Motley spoke with clarity and great dignity. Motley straight away acknowledged that the pledges by nations to cut emissions were leading to double the agreed safe limit of atmospheric greenhouse gases. Individual countries promising the most ambitious emissions reductions were embedding carbon removal technologies, not yet invented, into their proposals.

'With all these pledges, we can still only just stay beneath 2°C,' Mia Motley continued. 'This is at best reckless and at worst dangerous. On finance, we are $20 billion short of the $100 billion which has been promised, and this commitment, even then, may only be met in 2023. On adaptation, finance remains only at twenty-five per cent. Not the 50/50 split that was promised or needed, given the warming that has already taken place on this Earth. Climate finance for the front-line Small Island Developing States declined by twenty-five per cent in 2019.'

Motley then reminded the conference that: 'Failure to provide the critical finance, and that of Loss and Damage, is measured in lives and livelihoods in our communities.' This is what Dr Saleemul Huq had been saying over and over again in his efforts to further the cause of Loss and Damage, a funding mechanism whereby those nations (and corporations) responsible for the most emissions pay to help those who are not responsible and who suffer the most. The urgency for action on Loss and Damage payments increases as each year passes.

Motley pulled the three strands of emissions reduction, adaptation finance and Loss and Damage funding into one basket, stressing that the promise of the Paris Agreement would never be fulfilled if these three commitments were not honoured. Her words were directed beyond the boundaries of the conference hall, to the leaders of nations who, for political reasons, had not come to Glasgow. 'When will we leaders across the world address the pressing issues that are truly causing our people angst and worry?' she asked. 'Simply put, when will leaders lead?'

The vacuum created by lack of leadership over the entire period since the late 1980s to the present day is perfectly illustrated by Motley's clear articulation of the facts:

> The central banks of the wealthiest countries engaged in $25 trillion of quantitative easing in the last thirteen years. Twenty-five trillion! Of that, nine trillion was in the last eighteen months, to fight the pandemic. Had we used that twenty-five trillion to purchase bonds to finance the energy transition or the transition of how we eat or how we move ourselves in transport, we would now today be reaching that 1.5 degrees limit that is so vital to us.
>
> I say to you today in Glasgow that an annual increase in the Sustainable Development Initiative of $500 billion a year for twenty years, put in a trust to finance the transition, is the real gap, Secretary General, that we need to close. Not the $50 billion being proposed for adaptation. And if $500 billion sounds big to you, guess what? It is just two per cent of the $25 trillion!

Prime Minister Motley here highlights that the wealthy countries of the world are actively dismissing the Global South as collateral damage. The $100 billion promised to vulnerable nations over a decade ago has not been received. Motley clearly articulates that it is a tiny amount, yet countries like the UK, US and others in the EU, despite centuries of extracting resources and enslaving their people, among other colonial crimes, look the other way when countries like Barbados are faced with the dire consequences of our actions. Yet central banks have an endless pool of capital that can be summoned at will when threats arrive on our own shores.

A familiar COP feeling returned – that this was a waste of time, a greenwash expedition that had nothing to give back to all those who pinned their futures on it. I tried to dissipate my cynicism by taking a stroll around the pavilions.

Cry me a cryosphere

The vast exhibition hall with endless aisles of pavilions was teeming with masked faces, weaving in and out of different campaign launches, panel discussions, delegation meetings and so on. Up the wobbly stairs of the Bloomberg Philanthropies pavilion I was pleased to find a subsidised coffee bar where the servings were large and strong. From here, I continued to prowl.

Minutes later, I rounded a corner and bumped into glaciologist Professor Jason Box from the Geological Survey of Denmark. He was going to the Cryosphere Pavilion, or, more colloquially, the ice pavilion. Jason mentioned there were some interesting presentations coming up, so we made our way there after he was furnished with a near-lethal dose of Bloomberg coffee.

The accelerated melting of glaciers and ice sheets was the subject of the presentation commencing as we arrived, hosted by ETH Zurich, a respected research facility in Switzerland. I recall visiting this place back in 2014 and interviewing a Professor Knutti, following a tour of the supercomputer housed on the upper floors.

Four learned speakers delivered stark presentations highlighting the linkages between ice and sea level and the remaining time frames for adaptation of the world's cities. Other knock-on effects include impacts on weather systems, mass human displacement, migration, food shortages and increased conflict. While listening to these four experts speak about what is happening, the ridiculousness of the media hype in the main conference hall shrank to insignificance. This presenting space, with so much critical knowledge, had about twenty or thirty people in the audience watching and taking in what was being said. These scientists were deploying another lens on our changing world, bringing the crisis we face firmly into focus.

It was stated that the warming we are creating by not reducing our emissions means we are now committed to the loss of snow cover in the Alps, most likely to happen this century. The larger ice sheets and glaciers are also receding and sea levels are rising. The point is, we are now committed to a vastly different world, but the speed of that change is still within our grasp to alter. How fast ice sheets and glaciers take to melt is the difference between a managed adaptation and a painfully reactive, chaotic collapse of civilisation.

The waves of realisation were incredibly intense, but it was in the Q&A when the mood really sank. A person in the audience asked, 'Do you feel positive about the future of the cryosphere and do you feel you are getting the help you need to answer this emergency?'

Professor Jonathan Bamber replied, saying, 'The short answer is no,' before handing the microphone over to his colleague, Antarctic expert, Professor Rob DeConto.

Rob said, 'I am more of an optimist than Jonathan, you know…' He then seemed to stall awkwardly before saying, 'It is not too late, or otherwise, you are just saying it is not worth fighting, but we know that things look different in a +3°C world than they do in a +1.5°C world. So, there is a sense of urgency and that is why we are all here and that's why we took time out of our lives to come here and hammer on these messages. We need to convince policymakers to begin taking this seriously and make sure that what is happening in these negotiating rooms is pointing towards something that looks better than a +3°C future.'

As Rob passed the mic back along the line, Jonathan grabbed it again and set out to qualify his bleak 'no' answer, saying,

'Okay, I'll pass it on but I just want to… I said no but Rob is absolutely right. There is everything to play for.' Jonathan's voice echoed Rob's in being very stilted and uncertain, as he went on, 'We haven't lost. We can turn the dial round. We can move it in the right direction, but I've been going to these COP meetings for almost twenty years. The first one I went to was in 2003. I got a little bit jaded over time. Not much has happened in those twenty years. I am not saying it is not going to happen here, and Paris was a big step forward. It was a really important step in the right direction but we are such a long way from where we need to be!'

He then passed the mic onto Professor John Pomeroy, who said that given the carbon emissions reductions announced so far at the COP, we should be signing the death warrant of the iconic Peyto Glacier in the Canadian Rockies. He ended his contribution, saying, 'It's not a happy story, is it?'

It was actually Professor Matthias Huss from ETH who brought the conversation back round to a more positive momentum, saying that we should be realistic and optimistic because that is the only way we will save whatever we can that is left to save.

It is worth underscoring that the tone of gloom that pervaded this session was in large part due to the failure of the world leaders and their corporate entourages, who were continuing to ignore the existential consequences of failing to make structural changes. Failure is their choice and they are *still* choosing to fail. These academics can see it and realise the error and the pain to come. These are the true scientists' warnings but who is listening?

After the session, I caught up with Jonathan Bamber and asked him to explain what most worried him about the loss of the cryosphere. He pointed out that the total sea-level rise from losing both Greenland and Antarctica would be 65 metres

(200 feet). Just 10% of that, 6 metres (20 feet), he said would be 'unimaginably catastrophic for humanity'.

I wanted to know how we attach real-world consequences to more conceptual notions such as a global mean temperature rise of 1.5°C above the pre-industrial period (1850–1900). Bamber continued: 'If we stay below 1.5°C, I think we can limit the worst consequences of sea-level rise. And it's all about the rates; it's how fast sea level goes up. Because if the rate is low enough – at the minute, it's about four millimetres a year, which doesn't sound like a lot – we can adapt. If it's too fast, then adaptation is not really an option and we're going to see migration on a scale that we can't really imagine.'

It is the speed of change that is critical. I asked him what we can expect if we continue on the business-as-usual trajectory, towards 2–2.5°C. His opinion was: 'If we go with NDCs, or what we are currently doing, then things are going to look a lot worse and there's no way that we can adapt in a managed way. It is one of the most serious consequences of climate-change, and, yeah, the high-emission scenarios look really grim.'

Other reports in the scientific literature highlight the potential risk of sea-level rise this century up to two metres and some that could be as high as four metres. I asked Jonathan, what does this actually mean? He said, 'So two metres of sea-level rise would flood on an annual basis, based on current population characteristics, about 630 million people around the world. That is a tenth of the population of the planet. That in itself is unimaginable. If that happens, we are looking at the breakdown of civilisation as we know it. We are looking at global conflict. I mean, that is unthinkable.'

I returned to my own feelings on witnessing the big-name speakers in the next room and asked whether he considered the outlook from Glasgow to be hopeful. He said, 'That is a really difficult question. My honest answer is I am not overly optimistic because I've seen the negotiations. I've seen the processes. I've seen the rhetoric before. I don't see anything significantly different here.'

I suggested that if the world leaders were failing us, then perhaps we had to look for something else.

'Yeah,' he told me. 'I think there are glimpses of optimism. I think that the youth movement, the mobilisation of civil society, and their increasing concern and direct action, I think that demonstrates people are concerned. Then if governments become concerned, because governments want to stay in power, and they have to do what the people think is important, they have to respond to those drivers.'

The Cryosphere pavilion was preparing for an evening breakout of wine and nibbles but I didn't fancy that. My mind was spinning with all the input received over the last couple of hours. The one upside that I could see to all this was that my COP lens, through which I could get a measure of the proceedings, was now in place. There was a real climate emergency and there was a fake climate emergency; they co-existed at the same conference, with two very different sets of rules.

This split personality of the COP was amplified further when I passed through the exit to the security boundary. More than guarding the perimeter, the police appeared to be defending it. The crowds outside had thickened. There was tension, a raw energy in the air. Drummers pounded a constant beat and banners stretched out along walls, across the ground. I cut through, passing a mobile food counter offering free food to everyone. I stopped and had some rice with a vegan sauce. It was good. Looking around, it was undeniably peaceful, but the tension remained. It felt different from previous COPs. The energy was outside.

These were the people who were frightened by what is happening to the wider world. These were the people who see the lies and the greenwash. They were in every direction – the beats, the slogans, the chants. The climate march in Madrid on the Friday of the first week of COP25 had been like a wild carnival. Here at COP26 in Glasgow, it was more tenacious, focused. I had to wonder, where was this going?

Descendants of Scott

Hordes of delegates were bundling their way onto the COP buses going into town. It was too congested for me. Feeling anxious about COVID, I thought I would walk back to the hotel. It was a mild evening strolling along the banks of the Clyde, and I was back in twenty minutes.

I dropped all my belongings in my room and followed the call of my stomach, heading back downstairs to the bar to contemplate dinner. The hotel was relaxed about people bringing takeaways to eat in the lobby. I ordered a cold bottle of a beer that had been aged in whisky barrels. It had a pleasantly rich, tangy flavour, with a refreshing moreishness to it. It was gone in no time, so I ordered another. As I did so, my phone buzzed. It was Beckwith asking if I wanted to join him and two others for dinner in a nearby restaurant. They were in the main train station, so I finished my beer and walked across to meet them.

Beckwith looked in good shape, perhaps a little more portly than Madrid two years before, but so did I. He was with a couple, also from Canada, called Heidi and Charles, who were running a version of Scott's Climate Show. As we sat down to eat, they explained that in the wake of Scott's death, the Climate Show had continued but after some disagreements it had split into three groups. The three groups were continuing with Scott's press conferences in the similar show format, with press-conference slots donated by NGOs who have access to the facilities. Despite the split, they worked as a co-operative, collaborating and cross-promoting the shows. This was the Scott 'afterlife'.

Beckwith mentioned that there would be an event to commemorate Scott later in the week in the Blue Zone. Scott had had an impact and, as living proof, these three groups were keeping his legacy alive.

The evening was drawing to a close and my phone buzzed with a reminder that I needed to have an early night. Tomorrow heralded a crack-of-dawn start with John Kerry, Norman Foster, and a number of mayors from around the world at the City Chambers on George Square in Glasgow.

Two climate bros

I exited my bed with a lacklustre bounce at around 5.30 a.m. The invitation read: 'Climate Breakfast with Mayors – A dialogue with Norman Foster & John Kerry'.

After COP22 in Morocco, when Kerry gave his outgoing speech as US Secretary of State, he had appeared to have lain low during the Trump years and was subsequently picked by President Biden to hold the title of US Special Presidential Envoy for Climate. An impressive title, at least in length. Lord Foster is known to me as a designer of expensive wineries, also that he resigned from the House of Lords in 2010 after a new law banned non-resident peers and MPs who decline to pay British tax on incomes earned outside the UK. His last speech in the Lords was seven years prior to his resignation.

I walked briskly to the City Chambers and was ushered up some stairs to a large room set out with tables for the mayors and other guests, with the media dispersed along the wings. The stage was set for the Mayor of Glasgow, Susie Aitken, Kerry and Foster. I moved down the side of the room and set up my camera as close to the panel as possible. Assured that I had a good angle for filming, I followed my nose to the breakfast anteroom. It was in here where everyone was loading up on coffee, juice and other titbits. I spied a typical Scotttish cooked breakfast that everyone appeared to be overlooking. The meat looked extremely dubious, but they also had a vegan option. I inspected both and concluded that the meat version might look inedible, but coated in egg and fried bread, it would kick-start my biological system into action.

The women serving were very pleased at my decision, and as I retreated, finding a safe place to contain the outward burst of egg yolk, one of the mayors from a Latin country came over to ask me if it tasted good. 'Fabulous!' I assured him, as he proceeded to the counter to execute his own due diligence.

Following two more coffee reloads before we were all summoned to the main hall, I found myself at the tripod's edge, watching the circular tables fill up with dignitaries. The main guests were introduced by Glasgow's excited mayor.

This was nothing more than one of the showy satellite events of the COP. It kicked off with John and Norman joking that they were quite used to sitting together chatting because they are neighbours in Martha's Vineyard, a small island off Cape Cod in Massachusetts. When asked about the prospects for the COP, Kerry answered confidently that when we leave Glasgow, 'I think we're gonna have the greatest increase in ambition we've ever had. The real issue is going to be follow-up.'

The trouble we have now is that 'ambition' is developing into a synonym for 'lying', or 'fantasy', depending on the motive of the person using it. The fulfilment of the ambition has not materialised, and so the carbon budget is eaten up year on year and the problem, subsequently, is far worse every year.

The breakfast meeting dragged on, with mayors being able to pose questions to the esteemed guests. Kerry got a lucky break with a magic diplomatic tap on the shoulder, telling him he was needed for a matter of state. He was escorted from the room beneath a thundery applause.

Foster stayed on, but I took the opportunity to pack up and run for the door. As I crossed back through the centre of Glasgow, I mused on why it felt so disappointing. Wouldn't it have been better to invite the press to ask the mayors pertinent questions about large-scale city adaptation and resilience building? Kerry and Foster could then chip in from the sidelines if they had something to contribute.

Having stayed at Alex's on the Saturday night, I had noticed that when the wind blew hard, it rattled its way through the whole building, sending chills up my spine. This was a widespread problem across the UK, where building infrastructure is either extremely antiquated or built to very low standards. These problems are exacerbated today by the success of the construction lobby at deregulating the industry. Doing things on the cheap increases the vulnerability of the many households who struggle to stay warm each winter. Poor insulation drives up energy consumption and bills. A session on how we retrofit the ancient building stock of Glasgow and then extrapolate the successful programme across the UK's 28 million homes would have been far more interesting.

Mood dissection with Professor Kevin Anderson

Inside the Blue Zone, I revisited the Cryosphere pavilion en route for a blood-curdling coffee from the Bloomberg barista on the mezzanine. Emerging out of the throng of people, I saw the masked face of Professor Kevin Anderson from the Tyndall Centre at Manchester University. I had interviewed Kevin online a couple of times during the lockdown for my ClimateGenn podcast, so I knew he would be here.

After watching another fascinating talk in the Cryosphere pavilion, Kevin agreed to record an interview, offering some thoughts for what we might expect from this COP. As it was fresh on my mind, I relayed what Kerry had said that morning at the mayors' event about there being more ambition at this COP than any preceding one, but that we had to watch what happens after the COP. To Kerry's remarks, he answered:

We haven't finished this COP yet, so it's a bit advanced to be saying that there's going to be more ambition at this COP. I think the Paris COP, whereby we would hold emissions well below 2°C, and ideally aim for 1.5°C, that was hugely ambitious. This COP is really about the reduction rates that are needed to deliver on that 1.5°C, and when you listen to the announcements on net zero targets and other various targets of deforestation and so forth, these fall far, far short of our 1.5°C commitment. In fact, I would say that they fall short of our 2°C commitments as well. So I'm not witnessing the same level of ambition that John Kerry is interpreting from what's happening here within the COP.

Since arriving in Glasgow, there had a been a stream of emails from the UK Prime Minister's press office, listing achievements and stressing the commitment to 'keep 1.5°C alive'. The emails suggested a narrative for the COP that reminded me of Paris, that the leaders were busy solving this for us with photo ops and bold statements. The troubling aspect was that none of this addressed the levels of climate mitigation required to get us on a track towards 1.5°C or even 2°C. The cryosphere scientists stated clearly: there is no current action to slow the rate of climate breakdown. So the question is, why the perpetual rounds of misinformation?

Kevin Anderson said:

We have all bought into a narrative of we must not remove hope, we must not say it as it is, at least not in public. We must portray a jolly, cheery message that we can have win–win. We can have green growth. We can improve the quality of life for everyone. We can level up.

We can do all of these things and address this enormous challenge of climate-change, but our commitments around climate-change are 1.5°C and 2°C and they can be translated through the science into the rates of change that we require. These are completely different to this green growth rhetoric that dominates and echoes around COP26.

There is a growing disparity between the overly optimistic narrative that emerges here and the reality of what is not happening in the wider world. Polling in the UK shows that over 80% of the country believe we are in a climate emergency. As understanding grows as to what this actually means, then people's mood turns to cynicism and anger. Outside in Glasgow, it manifested in the people gathering at the gates, making a noise and drawing the attention of the media, as well as the organised civil society meetings across the city. They were looking at the COP and saying, 'Hang on, this is garbage.' We are missing the targets that define safety for our species as well as all others. I asked Kevin if the well of hope was dry.

Without pausing to think, he answered:

There are two ways to think about this. If you had asked me three years ago, do I have any real sense of where genuine hope, hope that is based on substance, where that might reside? I would have said that I am really struggling to see that. But now it has changed and that hope has come out, not from the leaders. In fact, they have been removing that hope. That hope has come from a sort of rag-tag of civil-society groups. From the youth movements, from the other civil society groups that are out there that historically would not have been the usual suspects.

Now, there are all sorts of people, from all sorts of backgrounds and many parts of the world, who are recognising the challenge. They are seeing the challenge of climate-change everywhere around them a lot of the time and then they're hearing rhetorical nonsense from our leaders. So I think there's a wide recognition now amongst many in civil society that we need to address this issue. There is some sort of coming together of a new sort of messy partnership that I think is asking questions that are simply not being addressed by our senior leadership.

The failure of our leadership was becoming so blatant that it was woven into the comments of those in high office who came offering hope. Even Kerry had admitted in his statement that we have all these pledges – 'more than we have ever seen before.' But once they leave, they forget. People are noticing the failure. Kevin continued:

That failure, because it is so clearly evident, many people in civil society are aware and realise that changes must be brought about. They are not sure how to do it but, in their various ways, they are engaging, and in my view, it is their various forms of engagement that are changing the tenor of the debate. Now, whether they can change the tenor of the debate quickly enough to respond to our 1.5°C challenges, we don't know. I think it is highly unlikely for 1.5°C. It's still possible, just about, for two degrees.

With climate-change denial now a very fringe position to hold in society, Kevin highlighted what he regarded as more pernicious: mitigation denial. It is more pernicious because those doing it are the very people who are meant to be guiding us through the crisis. In his words: 'This is often even carried out by climate scientists, by senior academics, by the senior politicians. They are denying the levels of emission reductions that are necessary, levels of mitigation necessary, to deliver on the commitments that we've made, that are informed by the science.'

We arrived at a point in the discussion where the complex characteristics of system change mean that we cannot easily foresee how or when significant change will occur, only that it will. Kevin continued:

When I hear and see at these events the Norman Fosters of this world, the Mark Carneys of this world, and lots of the other very senior people involved in climate-change, such as the people you see on all the screens here, these people capture the problem that we face: that this particular group – of which many of us aspire; academics are part of it at the fringes – I think encapsulates all that is wrong in our response to climate-change. We are unprepared to recognise that this is an issue of consumption. That the issue of consumption has been driven by a relatively small group in our society. They are spread across all the countries of the world, but they are very small parts of each country. They are also the obstacles to meaningful change.

He finished by saying, 'The physics doesn't care about legal niceties, eloquent speeches, or sharp suits. It only cares about the CO_2 molecules.'

COP26: Stop Your Bullshit!
On the Friday of the first week of the COP, a Youth and Public Empowerment Day march took place in Glasgow, ending in George Square. I was in the media centre when I got a message from Jason Box saying that he was attending the march. We met

in the Bavaria Brauhaus, a rather grand-looking pub north of the Central Station, from where we could catch the marchers as they entered the square.

I hurried along the banks of the Clyde in the damp afternoon to my hotel, dropping off my bags, continuing by the side of the train station, up the hill, along the aptly named Hope Street, towards the Bavaria Brauhaus. Inside, Box was just emptying his mini-stein mug of beer, so we ordered another quick round before walking around to hear the speeches.

Kevin made a point about civil society being more aligned with the science of climate-change than the political and wealthy elites who also descended on Glasgow. This was evident in the tones of those speaking. Greta Thunberg loudly declared COP26 a failure; she declared it had failed before we arrived. The crowd responded loudly, indignantly.

Thunberg relinquished the platform to other speakers from the Global South, those with stories of everyday casual suffering and worsening pain, the result of unrestrained consumerism, exploitation and resource extraction.

One of the last speakers to grace the stage with incredible presence was a woman from Namibia called Ina-Maria Shikongo. In Namibia, a Canadian company has done a deal with the government to extract fossil fuels from a pristine area of land called the Okavango. This example of contemporary colonialism typifies the crimes that are being inflicted on indigenous peoples around the world. That the company is Canadian casts my mind back to COP21 in Paris and the ridiculous charade by the shiny new Trudeau government, declaring 'Canada is back!', implying the dark days of the Harper administration were behind us. Resource pillage is thriving.

This is a shortened transcript of Ina-Maria's speech, reflecting the mood growing in the streets:

My name is Ina-Maria Shikongo. I come from Namibia. A beautiful country surrounded by two deserts, and today we have a company all the way from Canada wanting to frack up the Okavango, putting my people at risk. And I am asking COP26 today, why do you keep on whitewashing everything?

Let's talk about the genocide that you have imposed on our people. You came to our continent; you took our ancestors; you enslaved them. That was a genocide. You went to the Americas. You caused genocide there and you never talk about them. You went to Africa and you are still continuing with your genocide. Let's talk about Tambo. Let's talk about ReconAfrica. Let's talk about EACOP. What's happened to all of these new oil projects?

You people, we are tired of your neocolonial traits. We are here today because we know that COP26 won't do anything. They want to continue the massacre that they have been responsible for hundreds of years already. Your world is built on the blood of our people and we want change. We will not give up. We want justice for the genocide, for the ecocide that you have been causing.

You cannot keep on oppressing our people. You cannot keep on coming to our countries claiming that you are bringing development when all that you are bringing is devastation. You are poisoning our waters.

I am here to defend my homeland today. The Okavango is the birthplace of modern humanity. The oldest DNA is in the Okavango, and yet COP26 is here, whitewashing, cropping activists out of pictures. What's up with that? We shall defend ourselves. COP26, PLEASE STOP YOUR BULLSHIT!

The crowd cheered. Shikongo had captured them with a loaded speech, vibrant with emotion. This was the zeitgeist now. The championing of Paris Agreements and NDCs that would result in double the 'safe' global-warming level has lost its sheen. The people on the dying end of the deal are travelling the world, taking their cause to the COPs, to the G8s and G20s and the World Economic Forum.

Wherever they go, they are heard. People are sympathetic. Middle-class people in the UK and in Europe are not happy to have this blood on their hands. The young people who realise that climate-change has screwed their future are definitely not onboard with the renewed colonialism of the past. When we destroy their lands, extract their resources, poison their waters, and then pump the waste CO_2 into the atmosphere to jack up the global temperatures, it really pisses them off, and that mood of pissed-off-ness is spreading.

Professor Rupert Read, former spokesperson for Extinction Rebellion and, more recently, co-director of the Climate Majority Project, was here too. After searching for him in the crowd, I recorded an impromptu conversation between Jason and Rupert, intrigued to see how the ice scientist, tracking the collapse of the Greenland ice sheet, connected with the philosopher-activist, working to try to connect with the growing number of concerned citizens.

Jason began by saying, 'I saw really young kids marching with a look in their eyes that I haven't seen before. I was in the march and the majority of people were really quite young.'

This comment visibly resonated with Rupert. He replied, 'I'm very concerned. There's so many angry young people here, and they're right to be angry. I'm worried that this anger is just going to multiply if it doesn't find some kind of outlet or some kind of response. And anger alone is not usually the basis for success. You need to have anger tempered with love and with determination to act. So I think we are in a very difficult place. The kids are understandably furious, and they need something to help them to turn that fury into an action that can change their world for the better.'

Jason continued that he was pleased to see how civil society here were reading and assimilating the scientific research produced by him and his colleagues, literally using it to argue for change. He said that it is through science-based policy that we have a real chance.

Rupert responded, broadening Jason's view, saying, 'We need to be science based. We need to be ethics based. We need to be precaution based. There is a whole shake up of the curriculum that is needed. I think we in universities have to lead this. We have to begin with a more serious kind of curriculum change in what we teach, as well as what we research, and then we can be really credible in letting that spiral down into the school system as well.'

Jason continued, reiterating the threat of accelerating ice sheet loss that had been made by Professor Jonathan Bamber earlier. 'All we can really do is try to slow them down.'

Coming at the climate issue from a different lens, Rupert stressed his fears around the near-term shocks that can spiral out of control:

The thing that really keeps me up at night is food shortages caused by climate chaos effects. The potential for multiple simultaneous breadbasket failures. This could cut in way before you are getting any serious disruption from sea-level rise or even temperature rise. I get very frustrated when climate scientists say to me things like, 'Well, but you know, if we reach two degrees, it will be quite bad but that is a lot less bad than three degrees of warming.' Of course, that is true, but two degrees in terms of its chaotic effects could already be enough to bring down our civilisation.

Countries like this one, where we're standing right now, are dangerously complacent, thinking because we are so rich, we are going to be okay for the next generation. I am not convinced of that. So I say to anyone who is saying, 'Well, surely we in Britain need to take care of ourselves.' Alright, let's take care of ourselves in sensible ways. Let us look after our water supply here. Let us start creating real resilience. Let us start thinking about protecting ourselves and protecting others at the same time from what is coming, because what is coming is not going to be pretty but we can act now and still have some margin. We can act now to make it less un-pretty.

What Rupert was saying chimed with interviews I had recorded prior to COP. Experts confirmed that each year, our exposure to risk from droughts, floods, tidal surges, forest fires and other disasters is multiplying. When these extreme events start to interact with other events, society can rapidly break down. Governments and civil society have been unaware of the need to plan for adaptation and resilience building or unwilling to do so. Without such planning, we edge closer to system-wide chaos.

As the crowd dispersed from George Square, Professor Box signalled it was time to investigate a nearby whisky bar that my artist friend, Alex, had suggested we visit – The Pot Still, on Hope Street.

Scott's spirit at the COP
It was Saturday morning and I had much to do. I dragged my weary flesh into the media centre to edit some of the recordings of the day before. As thick, treacly coffee embedded itself on my palate, I continued to work at a pace.

It wasn't long after lunch when I got a message from Beckwith reminding me that the Scott tribute session was approaching.

The main Fridays for Future march was also scheduled for today but my to-do list was still long, and if anything dissuaded me from a long march across the city, it was the heavy rain pounding outside.

Crossing the main concourse of the Blue Zone, I bumped into Professor Peter Wadhams and his wife Pia. I had interviewed Peter many times over the years in Cambridge and other places about his extensive knowledge of polar ice. Peter and Pia were relaxing before a press-conference that involved one of Scott's Climate Show descendants. We had a quick catch-up before they had to go on, and I shot off to try to freshen up a bit. The fatigue of the last week was creeping in.

Jogging heavily along the corridor, I bumped into Beckwith, who was wearing his grey flannel suit and looking quite cheerful.

'All good?' I asked.

'Yeah, I think so. You missed the ceremony outside with Scott's ashes,' he said, enjoying the look of shock creep over my face.

'What? You had his ashes here?' I said, my eyes bulging.

'Yeah, not all of them. Just about a leg's worth or something. It was actually quite nice. We all met by the Clyde and gave him a send-off into the river.' He then faltered, looking down at his clothing. We both found ourselves looking at a large, ashen patch on his jacket.

'What is that?' I sensed I knew the answer.

'Yeah, the wind really got up today. When they threw the ashes, they blew back right across my jacket.' His calm Canadian demeanour was not fazed by having the weight of Scott upon him.

'So that is Scott?'

He nodded in the affirmative.

'I think I need a beer,' I said. 'Want one?'

'Yeah, sure!'

Of course, Scott had to get to Glasgow. It was the final outing. An hour later, we walked among the pavilions to find those gathered to commemorate the great warrior. Earlier in the week at some point when I was running along the COP corridors, I had stopped to speak with a large collective of Scott's acolytes, a decent and well-meaning bunch.

Even now, as I attended his final farewell, I found it hard not to think of Paris and successive COPs, where Scott had surpassed himself with his form of intensely surreal rollercoaster activism, openly describing himself as a holy knight and warrior, burdened by his large sword of responsibility to save us all. At times, I wasn't sure if it was all an elaborate work of performance art. He had gone, but what remained for me was a head full of comic sketches and mad zigzags up and down my emotional Richter scale. There was quite a gathering assembled here to give him a send-off, so his unorthodox style had had an impact. I took a moment to smile and to enjoy this last beer with Scott, dispersed among his followers in Glasgow.

Capping off Week 1

That night, I opened another of the bottles of wine from Bordeaux that had been delivered to Glasgow prior to my arrival. I chose the Chateau Brown, 2015, Pessac-Leognan, and made my way to the bar to ask if I could borrow a glass. The staff were, as ever, incredibly hospitable, offering a glass and seeing no objection to me tasting in the bar area. As I was waiting, a lady called Theona, from the Outer Hebrides, struck up a conversation. Theona was here to support the just transition for rural Scotland and, specifically, stand up for her community's interests in the midst of opposing forces, whilst the light was shining on Scotland. Theona had some good activist principals and suggested a long list of people that I could potentially interview if I were to throw my lot in with the Hebrides.

Instead of commitment, I offered her a glass of Chateau Brown, an estate, I am told, that has 55% of the land under vine with the rest given over to orchards of cherry, pear, plum and apple. The estate places emphasis on biodiversity and, as with other producers in the region that I have spoken to, they do not certify as organic because of the necessity of using synthetic inputs in extreme circumstances.

After a day of running around Glasgow, this wine made from Cabernet Sauvignon and Merlot, with a tiny addition of Petit Verdot, was quite sublime. The taste of cassis and brambly fruit, with a dense tannic structure, triggered a release of endorphins, calmed my aching legs and cast Theona's description of the Outer Hebrides into full Technicolor.

It's gloomy, but...
Typically, on the Sunday between the two weeks of the conference, I had no plans, as this was a day of rest. This year, I had arranged to meet Dr Paul Behrens in the hotel lounge to record an interview about his new book, *The Best of Times, The Worst of Times*. Paul is a bright young academic, full of good energy and enthusiasm that he applies to decipher the reams of live data and probabilistic outcomes that come together at his fingertips.

His book is an insightful and realistic assessment of potential routes to our uncertain future, and so it was a great pleasure to be able to speak with him and record an interview.

Paul is very good at using social and engineering data to pick out how counter-intuitively many of us view the world. There is a narrative that pervades in some quarters that the key to solving climate-change is by reducing the global population, which would naturally reduce consumption. This conversation advances towards how many people the Earth can carry and at what point do we overshoot its capacity. Many assessments say we are way over-populated and need to drastically reduce the global human population.

This is the bit where we all look around and say, 'You first!' The next part of this narrative is that the population rise is due to people in poorer parts of the world like Africa and Asia, and that now they are being lifted up out of poverty, their rising level of consumption added to *our* levels of consumption will collapse civilisation. Now we are combining elements of *us plus them*, as well as a healthy dose of fear, into the equation. This discussion can take a disturbingly racist and, I would assert, dishonest and savage character. Aside from this, it also hides some very important truths about our own societies, especially here in Europe, that should concern us.

The population of Europe is falling. Paul's data showed that by 2050, we will be losing 1.4 million working-age people per year, which will have a huge impact on the economics of Europe. Even now across France, Italy and Spain, for example, whole villages are derelict and abandoned. Schemes exist to encourage people to purchase properties for 1 euro with a view of establishing communities. Perhaps this might be the way forward when renewable energy, water-harvesting techniques, and fast internet will enable self-sufficient communities to flourish. We are some way from there.

The narratives of fear about invading migrants perceived as a major threat are very much the reverse. Europe and the UK are not full and could easily absorb migrants from the Middle East and Africa into our societies. Many of those people on the move are doing so because of conflict and extreme weather that have been quite often caused or exacerbated by politicians or corporations in Western countries. Even discounting any sense of responsibility, encouraging acceptance of other people around the world could help us build greater resilience, peace and a more secure future.

The armies we have trained to blow each other up could play much bigger roles as first responders in climate catastrophes, as well as helping to protect and restore nature. Instead of chaos, we could envision a richly biodiverse Earth system that restores and sustains life.

I said goodbye to Paul and walked across the city centre to the Babbity Bowster for a farewell lunch and ale with Professor Jason Box. Although the lunch was good, there was a real sense that the foregone failure of the COP to deliver on its political promises was getting into the psyche.

Afterwards, walking back through the streets, huge numbers of people expressed themselves through dance, music and costume. The camaraderie in the streets was taking back the agency from the disconnected politicos. Naomi Klein's Paris comment about not 'abdicating the power' was expressed in every action. We had all done that for decades, and in that time, those with the power had stolen or squandered as much as they could get their hands on and were now clueless as to what to do next.

The boomer party that had raged throughout the latter half of the last century and into this one has ended in disarray. There is a rising tide of people who can stomach no more. They have crossed a Rubicon and become activists. Even this term is something that can drive anxiety into a great many people. Activists are unpredictable and say uncomfortable things. Sometimes they get in the way with their protests and make life inconvenient. Also, the government hates them. Quite often, the media does too.

I recalled in Vienna in 2012 when I attended a talk given by Dr James Hansen when he was still the director of the NASA Goddard Institute. He said that when he submitted a scientific paper for review, he had added some suggested action at the end of it. The paper was refused for publication. When he asked why, he was told that it was because of the prescribed action. If you include this, then it makes you an activist. Such a thing was not acceptable. There are NASA scientists today who are among the most vocal climate activists in the world. One is Dr Peter Kalmus, who has built a huge social-media profile and takes to the streets with the US activist group Climate Defiance, calling out polluters and for an end to fossil-fuel production. This is how times have changed.

Peter was recently arrested for civil disobedience – as Hansen has been, as Jason Box has been, as Naomi Klein has been, among many others who believe that peacefully and non-violently transgressing the laws enacted to protect us all is the best way to stand up to the broken socio-economic system that poses the greatest risk to our wellbeing in the near term.

I had not considered myself an activist. It was only when someone thanked me for my activism that I even considered my interviews on climate-change to be more than a genuine quest for information. When I contemplated the implications of what I was being told by credible and multiple sources in many different disciplines, I realised that this posed a risk to all of us. Activists in recent years have become more extreme, and I believe this trend will continue until society responds to their signal – not with incarceration but with a decisive action plan to reorient our experience of life back within the confines of nature. The activists are a part of the complex system of human civilisation, emerging from the scientists' warnings having fallen on deaf ears. They are warning the rest of us that the ark is heading for the rocks. They cannot cease warning us until we are safe.

I said farewell to Box as he headed off back to Copenhagen and I headed back to my hotel room to prepare for the week ahead.

Adapt Now!

Cascades of multiple disasters, resembling a domino effect, are becoming more widely discussed by a wide range of researchers around the world. In 2021, I conducted a series of interviews to better understand how such a cascade could be triggered. The timing of shocks is unpredictable, but their forms, in terms of water shortages, broken supply chains, food market spikes or, in the future, maladaptation, are becoming more foreseeable.

After publishing the interviews as part of my ClimateGenn podcast series, I received a call from Pooran Desai OBE, founder of OnePlanet and an innovator and entrepreneur in the buildings and technology space. Pooran sounded stressed. He had just listened to a recent episode with Sir David King, discussing how the Arctic tipping point had crossed a threshold of irreversibility. This spelled trouble for stable weather patterns and a multitude of other Earth-system interactions.

After much discussion, we agreed to organise an event in Glasgow to explore concepts around adaptation. We worked with a group called After the Pandemic, based at the University of Strathclyde in Glasgow. I invited interviewees who we knew would be in town to participate in our event, titled 'Adapt Now!'

Professor Alice Hill, a former US judge and director of the US National Security Council during the Obama administration, working on pandemic response and climate-change risk and acting as a special advisor to President Obama, was one of the first people I spoke to. Alice had recently published her own book titled *The Fight for Climate after COVID-19*. It takes a pragmatic look at the components of building a resilient society. One of the early chapters that resonated is titled 'Leadership Matters', with a poignant look at how the UK government responded to the COVID crises. Alice breaks down the failures at the top of government that led to increased loss of life and the near breakdown of the National Health Service. Lessons learned from the pandemic could serve us well for the larger crises that climate experts are forecasting.

For over a quarter of a century, the COP process had focused on mitigation and achieved little. The failure to achieve real-world action has had consequences. Alice's message was clear: we need a comprehensive multilateral approach to adapt to climate impacts that can no longer be avoided.

Former UK Chief Government Scientist Sir David King was another interviewee analysing the steps that could lead to a breakdown of society. In June 2021, Sir Dave set up the Climate Crisis Advisory Group (CCAG), a group of global experts from science and policy. CCAG set about publishing reports and conducting open livestreamed briefings on a frequent basis, to keep pace with the changes unfolding in the fields of climate and biodiversity. This agile approach was to work in tandem with the longer, multi-year reporting of the IPCC. If we think back to COP21 in Paris, the late Dr Rajendra Pachauri said that the current IPCC reporting cycle was no longer appropriate. Formulating climate reports over the course of seven years, valuable though they may be, is the opposite of an emergency response.

One key focus for Sir Dave and CCAG has been the melting of Arctic sea ice covering the polar north. Researchers have demonstrated how the Arctic is connected to every region in the world via ocean and atmospheric currents. This change of state, from frozen ice cover to exposed dark ocean, has the potential to destabilise our weather systems, accelerate warming in Greenland and contribute to melting Antarctica, among many other linkages. In one CCAG briefing, a panel of the world's leading scientists explained how the Arctic had passed its tipping point of irreversible change. This is happening much faster than many scientists expected, at a global mean temperature warming of 1.1°C, not the 2°C+ scenarios that were previously forecast. This was a signal that Earth's sensitivity to warming is greater than previously thought.

Another one of my interviewees was a young activist from Namibia with Fridays for Future, called Jakapita Kandanga. When I first met Jakapita on a web call, I couldn't get my sound to work on my laptop. Thinking it may be an issue at her end, she tested all her plugs and settings. The issue went on for about ten minutes. In that time, I saw before me a young, smiling woman, who I might expect to be kicking back and enjoying the formative years of her youth. When I finally got my machine working, I realised that this young person possessed great strength and had a powerful voice, articulating with clarity the challenges she and her community are facing.

Jakapita explained how fossil-fuel companies are entering their lands in the Okavango Delta, arranging deals with corrupt government officials to prospect for resources. The Okavango is considered to be one of the most important wetlands in Africa, known for its rich biodiversity, home to a wide range of large mammals, including elephant, lion, leopard, cheetah, buffalo, hippopotamus, crocodile and hundreds of species of birds. Yet, their future safety, along with Jakapita's community, is disregarded in pursuit of profit.

Jakapita also explained how the worsening multi-year drought has made it harder for her family to keep livestock. Water shortages have led to sharing their animals with neighbours and trying to live with less. If the water supply is poisoned by the extractive industries, it will lead to growing desperation. Confronting all this, Jakapita travelled from Namibia, in south-Western Africa, to Glasgow, to be part of the Fridays for Future demand for change. It was impressive.

I asked Professor Kevin Anderson to join the panel. His work on how policymakers interact with the wealthiest 1%

of society, academia and civil society indicated that policy was being constrained because the real changes needed would mean reducing the consumption of the wealthy. Senior-level academics and policymakers, aspiring to join the elite groups, seldom spoke out about what their own research implied. This weakened the call for the critical policies required to rein in consumption. However, the research also suggested these dynamics could shift *if* people within the groups broke ranks and spoke truthfully about what the research was telling them. Kevin identified younger academics as being less driven by status and more inclined to speak out. Will those empowered to create system change realign before it is too late to avert global-scale social chaos?

The last interviewee I invited was Professor Saleemul Huq from the Independent University of Bangladesh in Dhaka. I had interviewed Saleem in Istanbul, London, Paris, Bonn and Poland, as well as several times in between. His knowledge and experience regarding what matters to climate-vulnerable nations like Bangladesh is among the best. He has long been calling for a Loss and Damage fund to be established for the benefit of those worst hit by climate impacts. This is a critical piece of the climate justice jigsaw that needs to be adopted worldwide, prioritising the most vulnerable. The vulnerable are not just in far-off places anymore. They are all over Europe, North America, Africa, Asia, stretching to the Arctic Circle and beyond. It is time for new collective thinking on this.

Although Saleem said he could make it, he was unable to attend at the last minute. I later saw him being given tea with the then First Minister of Scotland, Nicola Sturgeon. I hoped this meant he was progressing well in this city gathering of high-level tumbleweeds.

I hiked across a windy, overcast Glasgow to get to the University of Strathclyde, a complex of modern buildings on top of a steep hill. My ankles creaked walking up it. Inside the warm and colourful main building, I met Pooran with his assembled team and began drinking coffee. Panellists were due to arrive any time now and our confirmed one hundred guests were already taking refuge in the college café.

The event space was a large black-walled studio, with chairs arranged around circular tables facing the stage. The podium was placed between the panellists' table and the technician managing the livestream, all plugged in.

We wanted the discussion to be inclusive, with a high level of panellist-audience interaction. Guests filed in and quickly filled the tables. More chairs were put out and a line of people sat along the back of the room. I stood at the podium with Pooran as he welcomed everyone, giving the back story as to why we were here. He joined the panel and I began by introducing and asking each panellist a question. I then immediately opened the discussion up to the floor. The mood in the room was engaged and focused. Each opportunity to pose a question, or make a statement, was met by numerous hands shooting up in the air. We tried to get to everyone, but what follows is a collage of points that provides an impression of the topics covered.

An early question put to Kevin was whether we were right to be talking about adaptation at a time when mitigation is the focus of the negotiations. He said, 'We're absolutely right to be talking about adaptation and we should have been talking about it for the last thirty years. We need to put the infrastructure in that is suitable for a very different climate from the one you've got today.'

Alice Hill stressed one critical bit of information. For every $1 we spend now on adaptation, it equals a future saving of $11. This should be stated in all climate reports as they are put into the hands of policymakers.

Many of us use past experience, or historical records, to inform our planning for the future. Alice made it clear that in an era of consecutive and concurrent extreme climate events, the past is no longer a guide for what to expect from the future. We are in new territory. By over-relying on past experience, we actually increase our vulnerability to climate risk. She gave this example:

> In 2017 and 2018, we had devastating wildfires in California. The town of Paradise, a lower-middle-class town in the mountains of California, burned to the ground. Eight people died and 20,000 people overnight were thrown into another adjoining town called Chico, California. Chico already suffered from affordable housing challenges and suddenly 20,000 people appear with kindergarteners who need to get enrolled in school, people looking for jobs, people looking for a place to live. A highly difficult situation for both those that have been displaced and in the receiving communities.
>
> But the modelling doesn't really pick up what else happened. It turned out that the fires got so hot that the piping underneath Paradise melted, spreading toxic chemicals to the water supply. We then saw that the wildfire smoke caused health implications, such as increased asthma, increased respiratory disease, and spread all the way over to where I live in Washington, DC.

Alice continued, describing how a huge volume of rain in the atmosphere, known as an 'atmospheric river', fell on California. It landed on hardened land, scarred from the remnants of intense forest fires. The rainwater flowed over the surface, picking up debris, including large boulders, then ran down into reservoirs, contaminating water supplies. Alice said, 'We need to invest now in thinking through what the ramifications of climate-change are and make better decisions right now.'

Another question from the audience came from Dr Paul Behrens. Paul said that many discussions about climate adaptation are closed down, often by well-meaning people, because they see them as too doom-laden. The name 'doomer' is appended to anyone proposing investing in, or taking action on, adaptation. Narratives perceived as too negative are discouraged. Paul said, 'You need that sort of switch to happen in order for people to actually want to spend that dollar. So how do you balance this in your communication and how you speak about the challenges coming ahead, for adaptation?'

Sir Dave answered this question in the context of the insurance industry and, in particular, the reinsurance industry – the companies who insure the insurers. He stressed that as insurance industry losses grow due to extreme natural disasters, supercharged by climate heating, then the pressure to spend the $1 upfront to avoid the $11 later will be far greater.

Insurance is a subject that Alice Hill covered extensively in her book, and she responded, saying:

> Well, on the insurance point, we are close to a crisis in the United States with property and casualty insurance for wildfire. California is the fourth largest insurance market in the world. So insurers want to operate there, but for them, the fires in 2017 and '18 wiped out almost a quarter century worth of profit for the insurers insuring for wildfire. So at this point, some insurers have said, 'We don't want to insure property in areas that we know are at high fire risk.' What will happen is, eventually, this will even become too big for the reinsurers. The CEO of AXA says a 3°C to 4°C world is uninsurable for property and casualty insurance.

Alice's last point made Sir Dave's eyes bulge. He raised his hand to interject, pointing out that, 'Four degrees centigrade, forget it! We are never going to get a world of four degrees centigrade with anything like the humanity survival rate that we have today.'

Sir Dave continued with a cautionary tale of what is happening in Jakarta today:

This wonderful, bustling modern city, built out of the tiger economy of Indonesia over the last twenty years, will not be liveable within the next five to ten years because of frequent flooding. Indonesia is now talking about moving its capital to higher land. We are talking about a period when rice production in that part of the world collapses. Once you have been flooded with seawater, that is the end of rice production. It also includes south-east China. The flooding of paddy fields massively happening in just thirty years, unless we can manage sea-level rise.

So we are talking about very urgent problems. There will be no global economy like we know it today once rice production collapses. The global markets will shut down. They won't let the export of food to other parts of the world. We operate on a selfish basis. So we are going to see the collapse of the global economy well before we hit 4°C.

In the 2020s, as far as we know, we are in a moment of *being able to respond*. Work needs to be accelerated *now* on the social implications of planning for, as well as planning for after, these catastrophic events.

Alice gave an example of how one impacted region was able to deal with the complexity of having multiple communities as stakeholders:

The one way that is the most vivid to me is expensive, but involves deep engagement with communities. We have one example of that in the United States. The state of Louisiana along our Gulf Coast was hit with the BP oil spill. As a result, the state got a fair amount of money. It used that money to engage with a number of counties in that area.

They had thousands of meetings. This is a deeply Republican area in the United States, where climate-change isn't a particularly welcome discussion. But through those meetings, those counties got to where they actually identified the land that would disappear as a result of sea-level rise and subsidence. They're losing about a football field an hour of land in Louisiana. So it is important they figure out which communities will survive. Through that planning process, they were able to identify who is going to have to move.

Insurance for the poorer nations does not exist. It was a shame that Saleemul Huq was not here to discuss how a Loss and Damage fund, if adequately financed, could enable a just response to the increasing Loss and Damages that so many are suffering. It could also provide us with a model to be scaled up as Loss and Damage spreads across all nations.

After the catastrophic floods in Germany in 2021, Saleem was contacted by the NGOs working in the area to advise on disaster management. Lives were lost, property and possessions were destroyed. This was all too familiar to people in Bangladesh, who had been building resilience to worsening cyclonic and flooding events for years, actively saving lives. There is much knowledge to be shared on how to respond to devastating impacts. Also, if insurers decide their business is no longer viable, we will all be calling for Loss and Damage funds in the future.

Jakapita highlighted another issue that prevents Namibia's most vulnerable getting the urgent help they need. She said that the money that is given to the Namibian state seldom finds its way to the people who really need it. If the money is to reach those for whom it is intended, the nations giving money have to do the follow-up. Jakapita says, 'If the people really impacted by climate-change are not really benefiting from this money, then it is a useless deal for us.'

Alice added, 'One additional point on the question about Africa, it raises the migration issue. I focus on national security and climate-change, and here at the COP, for the very first time, that topic was featured. The Secretary General of NATO appeared and stated that his strategy for NATO was now examining the global security risks posed by climate-change. They are enormous. As we see water availability reduced, food insecurity increased, that all adds to the risk of conflict.'

Climate change as a threat multiplier reminded me of General Ghazi in Madrid. It has also become prevalent among military strategists regarding the pressures on global security arising from migrant hotspots. Alice described people who are forced to move out of desperation, from habitat and ecological destruction, as survival migrants, adding that, 'They're not climate refugees. They're looking for food. They're hungry. They need livelihoods. They need to be able to educate their children.'

The impact of these people moving is felt by those who are receiving them, as well as those who are losing them. Aside from any perceived security issues, it will require compassion and multilateral co-operation to manage humanely.

A member of the audience asked about the level of awareness of climate impacts in Jakapita's community in Namibia. She answered:

They live the climate crisis every day, but they do not understand, because they are not aware and they are not educated. The gap between our government and the people at the bottom is way too big. Our government attends COP and they sign these agreements, but then, when they go back to Namibia, they contradict their own policies. For example, our government signed the Paris Agreement, that included the reduction of carbon emissions, but then they're trying to invest in oil.

Jakapita explained that the Namibian people 'do not blame countries or companies that give out all these emissions. They blame God because they do not understand. I feel like a lot of attention should be given to creating awareness of what the climate crisis is. Then maybe people will be able to better adapt.'

There is a temptation to see this as different to how we are responding in countries like the UK. I would say it is, for many people, exactly the same, except we have resources to cope and are in a climate or we are wealthy enough, currently, to mask the impacts. The education gap in Namibia exists in many other places, too. If there is corruption in politics in Africa, that same corruption is at work in the UK and many other countries around the world, between corporate media, fossil-fuel industries and policymakers.

A question was posed to the panel, asking whether they thought this COP would succeed in moving us forward from Paris, or was it going to fail, in line with the COPs established pattern.

Kevin raised his hand for the microphone, and said, 'To me, this is why COP26 is bound to fail.' He pointed across the room to a large poster at the back.

The two words at the top of that poster over there: net zero. There was no reference to net zero in the IPCC Assessment Report 5. It's all over Assessment Report 6. The Committee on Climate Change in the UK, in its fifth Budget Report five years ago, it wasn't mentioned once, but in its sixth Budget Report, that same phrase is used between 3,000 and 5,000 times.

We use that expression now, and anything fits into it. Saudi Arabia Net Zero, BP Net Zero, Shell Net Zero. Every country basically aims at net zero. It has undermined the need for immediate and pressing action to a whole set of things that net zero catches.

Speaking about corporate greenwashing and corruption takes us back to the notion of agency and who is defining the narrative by which we choose to live. Do we buy into the lies that are cementing our road to destruction, or do we strive to reorient our lives toward something better?

Kevin said, 'I think we need multiple narratives, eclectic narratives, diverse narratives from different groups. Responding to climate-change is deeply cultural. So there isn't one message out there, it's an interplay, opening up to the sort of arts and social scientists, to narratives, to stories. These will not always be covered around carbon dioxide and climate-change.'

Reflecting on mechanisms for change and Kevin's comments, Sir Dave took another approach, focusing on our inherited patterns of behaviour across the Western world. He started with some key numbers. In 2000, the global population was six billion people. One billion were categorised as middle class, spending an estimated $10 to $100 per day, depending on their country of residence. Today, we are at eight billion population and the middle class is estimated to be three billion. Sir Dave said:

Now we see the global middle class rapidly expanding, far more quickly than the global population. We've got this very rapid rise in people who are all imitating our behaviour, and there is the big challenge. We can't have a planet of middle-class people, yet that is what we would all like to see, because we want to see them behaving like us.

That means we have to change our behaviour. We are the exemplars, and others are trying to imitate us. So if we are flying around in jets, if we're driving SUVs, if we are consuming the planet, we are setting an example to the emerging global middle class.

He went on to highlight the expansion of livestock and rice, which are driving land-use emissions of methane. We consume far too much meat for our health and for the health of the planet. This is the example we are giving to the rising middle classes around the world. He ended by saying, 'So there are all of these behaviour patterns, and that is a very big challenge to all of us.'

The ninety minutes ended, and the concentrated energy in the room was released. The session had focused on what must be done in the coming days, months, years. Adaptation is now a consequence of many decades of delay: dishonest mainstream media, corrupt politicians, dirty lobbying, lying corporations, and all of us who look the other way, saying it is too depressing to face.

Glasgow descends

The mood in Glasgow was not jubilant, as we saw in Paris, seven years ago. The people outside were furious with the crap that was going on in the heavily securitised 'ghettoed Blue Zone'. Inside the ghetto, I was stalking around, looking for an update on COP developments. My phone buzzed with a message from Professor Hugh Hunt from Cambridge with a single word message: 'Beer?'

Perhaps beer was the answer I was looking for. I met with Hugh later that evening and said I was trying to find someone to do a round-up. He suggested an academic from Cambridge, Dr Natalie Jones, who was inside the negotiations, observing the minutiae of the proceedings. Following an exchange of rapid texting, he arranged for us to meet in the hotel lounge the next morning to record a short overview.

Natalie arrived promptly at 8.00 a.m. Hugh was organising coffee. I was assembling my filming equipment. The hotel lounge had the familiar morning calm of many city hotels, designed for business meetings, hot drinks, pastries and enough nondescript music to shield us from background noise. Natalie was pushed for time as she was live-reporting from the negotiations and needed to get down to the COP. It was also nearing the end of the two weeks and she rolled her eyes when I asked if the pace was taking its toll.

We jumped straight in, asking Natalie what she was seeing in terms of commitments on emissions. She began by showing how the United Nations Environment Programme (UNEP) Emissions Gap Report had stated emissions were still way too high. She placed her hand low and moved it along a steeply rising axis to demonstrate how the NDC pledges kept us on track to miss the Paris Agreement commitments by a wide margin. To make the point clearer, she started the process again, this time trending her hand in a long downward slope, saying, 'If we were going to limit our warming to 1.5°C, we would be going down like that.'

The difference between the two was not unexpected. Natalie said, 'So I think next year, it will be interesting to see what the next iteration of that gap looks like.'

Interesting, but with emissions set to keep rising and extreme weather impacts already hitting us, increasing suffering around the world, there comes a point when the world stops believing in the pledges at all. Natalie went on to highlight the problem: 'Maybe we've closed it a bit with these pledges, but then it's still about the implementation gap. Even the countries who have the most ambitious pledges, for instance, in the UK, the Committee on Climate Change, have been saying, "Well, the UK has this great number of sixty-eight by 2030," but actually, the committee for the independent body is saying, "Well, where are the policies and plans to meet that?"'

They simply don't exist. The decay in our society allows these claims to be made by policymakers, despite them having no intentions of substantiating them. Natalie continues, 'Then on the other [hand], just a couple of weeks ago, was released the Production Gap Report, which is the other side of this. And you also have countries which have very ambitious net-zero plans, but all of their policies would be increasing the amount of coal, oil and gas they are extracting out of the ground. Those things are fundamentally incompatible.'

The miracle of 'net zero' strikes again.

As the minutes and hours dragged on, many people I spoke to were looking more and more hangdog. Agreements were being watered down. Confident voices on reducing carbon emissions, ending coal and accelerating the transition to green energy sounded weaker, and then completely absent. John Kerry's optimism from the previous week did not stand the test of even a short time.

A lot of the conversations that people were having seemed to be focused on whether it was worth having COPs any more. If the repeated outcome is to constantly delay meaningful action, while the available time to act ebbs away and emissions continue to rise, then a new approach is urgently required.

Wrapping up
As the COP wrapped up, my modus operandi was to collate perspectives. Some were captured in the dying hours of the COP and others in the days that followed.

The first voice is Kevin Anderson, who I bumped into in the Blue Zone. He agreed to comment on the draft agreement that had now been released prior to the official signing. It was, at this point, still subject to change.

He said:

There is nothing in there. I mean, it is an empty, vacuous document that is, no doubt, very carefully crafted by some wordsmiths, but it's full of 'urge' and 'we acknowledged the problems', and 'we regret that we haven't managed to do something in the past'. But there's nothing in there that compels action, and the small, very weak pointers that are in there are completely out of sync with the scale of the challenge that we are committed to.

I think it is worth just reiterating the challenge is not what we faced when we had the Paris Agreement in 2015. We are six years on; that is a quarter of a trillion tonnes of carbon dioxide that we have put in the atmosphere that will be changing the climate for hundreds, if not thousands, of years. So the challenge at the end of 2021 in Glasgow is not the same challenge as it was at the end of 2015. It is a much more difficult situation because we have again chosen an additional six years of failure, or six years of listening to the Kerrys of this world, the Obamas of this world, to the leaders who have demonstrated their contempt for the science and, arguably, their disregard for a lot of their citizens.

We discussed the different faces of the COP; the political shambles is one side but, especially in Glasgow, civil society groups were actively organising meetings, debates and even a People's Summit. Kevin had attended several events there and said, 'Quite a lot of the discussion is about: why are we in this position? What is it that got us here? How do we get out of that?'

This view corresponded to what I was seeing. The civil society spaces outside the Blue Zone in Glasgow were much more dynamic. My reading of the Blue Zone was one of impotence – by design. The theatre of world leaders and celebrities, underwritten by big polluters and profiteers. Segregated around in different halls were the indigenous peoples, the scientific experts and others, becoming increasingly distraught at the pointlessness of being inside there. The mainstream media remained focused on reporting a narrative that was all about hope and nail-biting negotiations where all is to play for.

The civil society groups and protestors had moved on. They were self-organising, searching for new narratives to forge pathways through the roadblock of inaction. Multilateral groups like Fridays for Future are critical because of their commitment to sharing their platform with people from those countries who have no choice but to rely on the COP progress. When dealing with the US, EU, China, Japan or any other large power, the most vulnerable nations lack the muscle to get a fair deal.

Prime Minister Modi said that India will continue to burn coal despite the climate threat, unless the wealthy nations provide $1 trillion to enable them to decarbonise at an accelerated pace. India is responsible for a tiny percentage of historical emissions, so they have a much stronger case for continuing to burn coal while they lift their population out of poverty. However, recent and ongoing heatwaves, attributed to human-caused climate-change, in India and Pakistan have been dire.

I asked Sir Dave King what he thought about Modi's $1 trillion demand, and he restated that the $100 billion a year the developed nations have promised to aid the vulnerable nations in reducing emissions and adaptation costs has not fully materialised. The total, for all the vulnerable developing nations, not just India, is miniscule. So far, the total that has materialised is $22 billion a year of public money. This has been topped up with funds and loans that are not always aimed at tackling climate-change and still only take the total to $90 billion.

Sir Dave said:

So we, the developed world, have not delivered. There's a real lack of trust. That's the background to Modi's well-placed comment about a trillion dollars. He is not saying, in a year. He is talking about the next fifteen to twenty years and it is a reasonable request. Are we able to afford it?

No question; we can't afford not to. It's going to spread around the world. There is no country that is not going to feel altered by climate-change events. I think it not just behoves us all to work together on this, it is necessary for our human survival.

I asked why we are unable to restore trust and push harder for the outcomes we need. Sir Dave responded, saying:

There's a very big problem, which is the power of the lobby system, particularly in the United States. These lobbies are so powerful that the majority view of the people in the United States is ignored by senators and congressmen, because enough of them are in the pockets of these lobbyists.

I spent six months working in Washington, DC and was simply amazed to find the number of lobbyists in that city and the way they make a living by persuading senators and Congress not to vote for climate policy. We have to see that the lobbyists are put in their place, and the only way to put the lobbyists in their place is to emphasise that the heatwaves happening today in Pakistan and India are going to spread around the world. Nobody is going to escape from the consequences of this. We really have to sit on these climate sceptics who have no respect for the truth represented by science.

In a recent interview with a retired US lieutenant general about the risk posed by climate to the national security of the United States, I asked him, 'Are we at a point when the United States is a security threat to the rest of the world?'

To my horror, without hesitation, he replied, 'Absolutely.'

Back in Glasgow, sitting in my stuffy booth in the media centre, up to my eyeballs in all this interview footage, my phone buzzed with a message from Beckwith asking if I wanted to participate in a round-up event. Professor Rupert Read was also taking part, as well as a virtual appearance from Dr Peter Carter, whom I had last seen in Madrid.

Everything I was hearing pointed to a COP confidence crash. UN Secretary-General Guterres had already said to the press that he did not think this COP would yield the *pledges* that would be sufficient to hold global temperature rises to 1.5°C. In an effort to preserve the mirage of optimism, Guterres made a peculiar statement about 1.5°C being alive but on a life-support machine.

I strolled the blue-carpeted corridors towards the Lulworth Press Conference Room to meet the others. A small crowd had gathered. The room was set up and ready to go. Peter Carter burst in on the screens and derided the UNFCCC process and its flagrant failure. I followed suit, then Beckwith said his piece.

Lastly, Rupert Read spoke, raising the questions that I had been asking others about the COP being fit for purpose, saying:

We can point our fingers at the fossil-fuel industry and the countries failing to show leadership, and there are other villains as well, but what about our own role in this? Does it actually make sense to be part of a process which is so direly off the pace? Are those of us who have been strong critics of this COP, of the COP process, of the recalcitrance of those countries, and so forth, are we nevertheless, in some sense, legitimating the process by way of being here?

I'm not honestly sure what the answer to this question is, but it seems to me that we have reached the point where we need to ask a question like that. We have reached a point where we need to no longer assume that taking part in this can-kicking exercise actually makes sense.

My second uncomfortable question is, should there be further elements to this can-kicking exercise? Should we all be assuming that, of course, there will be a COP27, and then a COP28, and so forth? Is that also now potentially part of the problem? Rather than of the solution? Should there be more COPs?

These are questions I had asked myself after COP22 in Marrakech, when the shock of the US election delivered a dangerous, malignant narcissist to the presidency. It was the anxiety that the true cost of these diversions into lunacy was the loss of time we have to act collectively.

Paradoxically, it can be argued that Trump has been a catalyst for a societal wake-up. When people realised he was hell-bent on destroying the global commons to suit his ego, they began standing up to him. People started to fear that they had things to lose.

'The vulnerable countries left Glasgow with tears in their eyes!' Dr Saleemul Huq

We do not have the leadership yet that will get us through the climate problem. COP26 was a terrible failure. After COP22, I reluctantly attended COP23 BonnFiji and found the raw energy of the young a signal of the emergent change we need. Now, in the wake of the failure of political leadership in Glasgow, the future again looks uncertain.

The last six COPs have all been in Europe, except for COP22 in Morocco. It has been easy for Europeans to come and witness the evolution of an emerging momentum for change. COP27 in Egypt is being dubbed 'the African COP', raising hopes that climate justice will come closer to being realised. That said, the Egyptian authorities sent early signals that protesting in the style we had seen in Glasgow was not going to happen. As we follow the COP into authoritarian territories, are we sacrificing the agency of the public to vigorously call out the pervasive empty rhetoric?

10

COP27, Egypt, Sharm El-Vegas

megaCOP, 2022

The plane flew along the Nile towards Cairo before descending over the Sinai Peninsula. In the late afternoon sun, the barren mountains and the red desert were dreamy. We tracked along the coast towards Sharm. An impression of human dwellings appeared like small geometric patterns on an alien landscape, and then, in one eyeful, I could see the flat sprawl of the COP itself: a series of temporary hangars, the permanent buildings of the Lamborghini Conference Centre, and a spray of miniature people casting tiny shadows in the declining light.

I left the airport terminal, navigating a network of tarmac strips pressed into the sandy surrounds. A young woman led me to a waiting people carrier, and, taking the last seat available behind the driver, I proceeded towards the fray of week two of COP27.

I had travelled to Egypt from Venice, Italy, with a moderately sized rucksack because I had damaged my Achilles tendon a few weeks before. Despite the lighter luggage, the ongoing strain meant my walking was getting worse. Entering the COP, I hobbled through security to the registration booths, then, with my badge, into a central piazza with a Hard Rock Café franchise at its centre. After dumping my stuff in the storeroom of the media centre, I headed back to the piazza to get my bearings. My bird's-eye view from the plane had shown that this was just a small enclave of the Blue Zone.

At the front of the Hard Rock queue, about to pay a ludicrous sum for a burger, was Beckwith. A year on from Glasgow, he looked the same as he greeted me with a big smile. The aroma of American burgers wafting across from the pop-up booth reminded me I needed to eat, but I fought it and let Beckwith enjoy his meal. He then took me on a brief tour, heading to the Cryosphere Pavilion, where endless talks on the diminishing rate and volume of global ice were being explained to a flowing tide of incoming and outgoing delegates.

He mentioned that the COP had a tendency to run out of food and water. On one occasion, the sewage system overflowed, causing a torrent of human waste to run freely down one of the main connecting walkways. He recorded it on his phone, and the sight of people crossing ankle-deep pee in the warm desert afternoon had a surreality to it that befitted a COP.

Noticing the pain I was having walking, Beckwith offered to carry my bags to my hotel. We left on one of the free buses laid on by our Egyptian hosts. The bus pulled out onto a long, broad stretch of tarmac, flanked by sand-swept, semi-constructed housing. We headed south towards Naama Bay, where my hotel was located. Beckwith said the place was like Vegas with its penchant for building replicas of monuments and cultural icons. We passed Hollywood with a few dinosaurs out in the front, followed by a huge casino with a Sphinx and some other faux-monuments. If we had carried on, we would have arrived in Old Sharm, where the dreamy Mustafa Mosque stirs the orientalist spirit but is, in fact, a modern construction. The old market resembles a recently developed shopping precinct more than a typical desert souk.

After checking in to my hotel, we went to grab dinner. I needed to sleep, but I also needed to eat. Ten minutes' bus ride down the road we found a small grill where locals were eating. We ordered a huge spread of grilled vegetables and half a chicken each with kefir and water, leaving no empty space on the table. I don't eat a great deal of chicken these days, so the aroma and sizzle that emanated from the grill was tantalising. For the brief duration of that meal, we were outside the hallucinatory Sharm El Vegas, amid local civilisation.

In conversing with Beckwith and others, it became apparent that the parallels with Las Vegas are not limited to sources of entertainment. Vegas is famously running out of water. It draws 90% of its water from the Colorado River, which is experiencing the worst drought in its recorded history. Lake Mead, from where Vegas sources most of its drinking water, is dangerously low and new restrictions have come into force to limit usage.

In Sharm, the cost of energy, powering the desalination of water, has fallen dramatically. Across Egypt, water scarcity for the population of 105 million people is a growing problem. The government is financing development of desalination plants, with an eventual water-production capacity of 8.8 million cubic metres per day, powered by wind and solar energy.

Much has been made in the press about Egypt allowing Coca-Cola, widely accused of being the largest plastic polluter in the world, to be the lead sponsor at the COP. The water sponsor, Nestlé, resorted to giving away thousands of cartons of water to overcome shortages, despite every delegate being given a reusable water bottle. The waste must have been colossal. After the water ran out, Coca-Cola announced that bottles of their flagship product and Sprite would be given away for free.

The one thing I struggled to find anywhere inside the premises was fruit. My only success was on the nuclear-energy stand where they were giving away bananas. I was told that bananas are good for protecting the thyroid against radiation poisoning. The small baby bananas were a rare elixir, as I consumed three immediately on discovery.

The vast layout of the COP was like a maze. I continually found myself lost and, having limited walking ability, missed several scheduled appointments during the week. The signage was tiny to nonexistent. Given that these huge events take place every year, the UNFCCC could at least have offered guidance to the host country about what works and what does not.

Gripes to the side, the morning sunlight tingled on my face as I gulped my coffee on the hotel terrace. The dining area looked out across a bulb-shaped swimming pool, surrounded by loungers and a sheer rock face that rose out of nowhere and had communication towers on top of it. The effect was asteroid-crater chic, and I appreciated it for a few moments at breakfast before hobbling off to catch one of the frequent COP buses.

It was around 28 to 30°C each day with low humidity. Palm trees appearing at intervals offered a little desert flavour to the proceedings. After a good night's sleep, it was time to go to work.

COP knocking – boycott or not?

In my time covering the COPs, except for Morocco, they have all been in Europe, which meant, naturally, they were very well attended by Europeans. By the time I touched down in Sharm El-Sheikh, the energy at the COPs had changed. In Glasgow, civil society was rejecting the processes and was busy self-organising to create a better future for itself outside the Blue Zone, in public spaces. I heard many say they would not attend COP27, and, though this seemed apparent on the activist side, any deficit appeared to be filled by youths from across Africa and the countries of the Global South.

As the COP drew nearer, there were many stories emerging about hotels costing thousands of pounds per week, with fossil-fuel companies and petrostates polluting the agenda. The hotel bit was not true. I can assure anyone that I paid nothing close to that and there were still rooms available at very close proximity to the COP. As for the fossil-fuel lobby in the negotiations, this was the COP where the industry that caused the crisis was represented by over 600 delegates, more than the combined number of delegates representing the ten countries most impacted by climate-change.

My reason for attending was because this was not a Western COP but an Egyptian COP, an African COP, and a vulnerable nation's COP. There would be plenty to disagree with as always, but there would be a new energy that, I felt, needed us to show up and demonstrate solidarity.

When I caught up with Dr Saleemul Huq from Bangladesh, we discussed what makes the COPs so essential to the most vulnerable nations. He said:

> Speaking now on behalf of the vulnerable countries, this is the only forum where we have a seat at the table to talk to the big guys. At the Security Council, we don't have a seat. In the G7, even in the G20, we don't have a seat. The UNFCCC is the only place where we have a seat at the table and can come and talk to them and hope to persuade them. We don't always win but sometimes we do. Like we got the 1.5°C in Paris. That was our doing; we persuaded everybody to agree to that.

The attention the COP was now attracting meant that pressure seeped into the negotiations. 40,000 people attended COP26 in Glasgow, 5,000 fewer than in Paris, but given the COVID restrictions, it was a very high turn-out. Paris and Glasgow were categorised as megaCOPs, marking the collective will of humanity to change course. Sharm El-Sheikh was not billed as a megaCOP. Expectations were that it would be one of the intermediary COPs, dotting the years between the megaCOPs. However, this presumed non-event pulled in 35,000 delegates, sending the message that the era of climate concern, lofty rhetoric and some action is upon us. Every COP going forward was going to be a mega-jamboree, a circus, a focal point for young activists and people with a future agenda, both good and bad.

I asked Saleem if he thought this growing jamboree-style COP could send a bad message, considering emissions continue to go up and people ask, 'What is everyone doing here?'

'I think we add pressure to the talks going on behind closed doors, certainly for the more democratic governments who have to listen to their citizens,' Saleem told me. 'The other positive thing that we get at the COP is a huge amount of media attention. Except for once a year, the global media doesn't pay attention. They come here and they want to know what is happening.'

Despite the polarisation of interests in the official COP between Global North and Global South countries, I was grateful to speak to two very clear thinkers from both sides' perspectives. Saleemul Huq always offers nuanced and thoughtful observations through the Global South lens. In the past, he has expressed fury at the behaviour of countries such as the US. The opposite perspective, coming from the US, is Professor Dan Bodansky from Arizona State University (ASU). Dan has a wealth of experience and insight from his side of the fence. What interested me was how both their views about the COP itself were almost identical. Both told the story of two COPs: the official proceedings and the more action-orientated side of the COP. They both clearly understood the soft power influence the latter could exert over the former.

One big issue that became the main talking point in Sharm was the acceptance of the Loss and Damage issue into the talks. Saleem had been pushing on this for years, and that it was on the agenda at all seemed to catch many by surprise. Dan was explaining how, if every COP is a megaCOP, we are going to see a lot more disappointment when big outcomes don't appear. At this COP, he was genuinely surprised about the emergence of Loss and Damage as a key outcome of the official process, as he explained:

> If you'd asked me a year ago whether this COP was going to establish a fund to address Loss and Damage, I would have said no way. But the Alliance of Small Island States (AOSIS) put it on the table in June at the intercessional meetings, and it looks pretty likely that the COP is going to adopt a decision addressing financing on Loss and Damage, including establishing a new fund. So that's one of the official outcomes from the meeting. That's a pretty significant development, addressing the huge concerns of Least Developed Countries and other vulnerable countries about the losses they're suffering from climate-change.

As good as the news was on the Loss and Damage funding mechanism, it is prudent to keep an eye out for the counterbalance narrative. Loss and Damage was a justice and compensation issue whereby big emitters would help low emitters cope with the damage caused by their behaviour. We in the Global North have sidestepped the issue of Loss and Damage for many years, despite the enormous damage already caused. Now, with a combination of mounting public pressure and sustained calls from the vulnerable nations, we are finally ready to advance this issue.

A jigsaw of emerging narratives
There has been much discussion around what scientists and climate experts are prepared to say on the record, as opposed

to what they really think off the record. The best experience of this I have encountered was watching an eminent person give a talk in Cambridge. This person was optimistic that technology would play a big role in fixing the climate problem. This was in the area of energy efficiency and carbon-dioxide removal from the atmosphere. Post-lecture, a few of us met for a drink in the pub. I then accompanied the speaker by train back to London.

During the forty-minute journey, we talked about various options for removing the carbon and the actual volume of carbon dioxide that was still being emitted. There were no signs of emissions reducing, despite the signals that environmental tipping points were being crossed. In a gesture of despair, the person put their head in their hands, crying, 'We're so fucked!' It is troubling when people agree to go onstage and tell one story of hope that, deep down, they fear is not true. It implies that knowing the truth is a privilege that others must be protected from. Not great when we need society-wide discussion and debate on how to move forward. Also, consider the personal toll of such a split between internal feelings and external demeanour.

I heard rumours inside the COP of the need to drop the language of 1.5°C that is the goal of the Paris Agreement (2°C is the official limit but to stay as close as possible to 1.5°C is the goal), and which was reaffirmed at COP24 in Poland when scientists presented evidence in a Special Report that going above 1.5°C risks crossing irreversible tipping points. We are currently on track to head somewhere between 2 and 3°C because we are not decarbonising our energy and agricultural systems.

If anything has changed in the post-COP21 Paris era, it is that 2°C is widely perceived as the new 4°C and 1.5°C is the new 2°C. The unprecedented climate destruction we are witnessing worldwide has taken many scientists by surprise, indicating the

Earth has a higher degree of sensitivity to temperature and that tipping points are kicking in earlier than anticipated.

After COP26 in Glasgow when the British COP President, Alok Sharma, stated that 1.5°C is alive but only just, it was seen as a controversial statement. Many said 1.5°C is dead and they should admit it, but many scientists vehemently stated we should continue to use it as the official 'safe limit' to stay within, even if we are likely to breach it. This has enormous repercussions that reach deep into all our lives, both in the Global North and the Global South.

I tuned into a live web stream in the media centre to listen to Professor Katharine Hayhoe, the well-known chief scientist at The Nature Conservancy, an environmental organisation operating across the US and seventy-nine other countries. Katharine was responding to news that there were voices in the negotiations that were calling for a loosening of the language around 1.5°C, shifting it toward the 2°C boundary. Katharine was incensed and reiterated that we must keep the limit and not let our ambition slip.

En route to Sharm, I had arranged to speak with Katharine, and watching this clip gave me the framing for my interview. She arrived at the media centre as arranged and followed me up to the first floor, where one of the best views from the COP could be enjoyed. Behind us was a panoramic view of the desert and foothills of the southern Sinai mountains. We located ourselves in the corner where a short while earlier I had seen former Irish President Mary Robinson interviewed in the same spot. As soon as they had finished, I moved in.

Katharine was seated, with microphone appended and ready to roll, so I asked her if the 1.5°C framing would indeed die. 'There's been a lot of discussion about exactly that all week because we're getting closer to the threshold that's needed to reduce emissions to a level that can ensure we meet that goal,' she told me. 'And it is a limit, not a goal. We don't want to go beyond 1.5°C; we want to be below it. If we scientists were to set a target, our target would be zero.'

Katharine then set out the context of where the COP process had brought us. The figures quoted below are based on the sum of the commitments of individual countries' pledges (the Nationally Determined Contributions, or NDCs, referred to earlier) to decarbonise and so reduce emissions:

So eight years ago, before the Paris Agreement, we were headed to a 4 to 5°C warmer world by the end of the century – four to five degrees! Now, according to the latest analyses on what countries have promised, we are headed to a world that is somewhere around two and a half degrees warmer than it was in the past. That's still too much, but it's a lot better than we had eight years ago. So we need to do as much as we can, as hard as we can. And if we do everything we can to meet 1.5°C and we get to 1.6°C instead, that's a lot better than if we just say, 'Oh, let's just aim for 2°C instead', and we end up at 2.5°C.

After the filming, Katharine vanished into the desert haze as I returned to my desk to process the interviews, while at intervals taking in the vast sweep of the Sinai view. Later in the day, I walked across the busy central piazza towards the pavilions. An area to the left was crowded with protestors chanting their rebukes and calling for the expulsion of the polluters. A large gathering surrounded them, clapping and filming. The authoritarian nature of the government here meant that protests outside were largely suppressed. There were stories of arrests and harassment. In my glazed state, it all seemed to pass me by.

Seeing the protestors inside the COP, they appeared sanitised and dumbed down. The performance was symbolic; in reality, their voices could not be heard. The right to protest is a vital component of a healthy society. It is part of the process of self-reflection, a signal that it is time to listen and change our ways. Stifling protestors is stifling the warning system, ensuring that the messages that are broadcast are sanctioned and cleansed. Social systems, like climate systems, have tipping points, thresholds and feedbacks. This kind of sanitisation I am witnessing here will be echoed by a more extreme response elsewhere.

Inside the exhibition halls, the easy switch from sanctioned protest to corporate trade fair was all too slick. I walked, still in pain from my injured leg, past the Egyptian megapavilion, through the rear exit and crossed the concourse to an equally large exhibition hall behind. I stopped for a coffee at a comfortable-looking pavilion called the Resilience Hub. While sitting, I checked my email and, by an uncanny coincidence, I had a message from Saleem saying to meet on this same pavilion to record an update on the COP.

I located him and asked for his perspective on keeping the 1.5°C goal. He gave me this answer: 'So the ability of the global community to stay below 1.5°C, which we've all promised to do, collectively, in the Paris Agreement, becomes more and more difficult by the day, and there is an attempt by some countries to give up on the goal. We certainly do not want to do that. The vulnerable developing countries want to stick to the goal. We all agreed to do it.'

Saleem emphasised that the Loss and Damage issue is not just for the Global South but for everyone. At the current 1.1°C above the pre-industrial baseline, we are seeing suffering from Loss and Damage everywhere. He went on to say, '1.5°C is itself going to cause a lot of damage. 1.51 will add more. 1.52 will add more. So every fraction counts, and we need to be holding it as much as possible to 1.5°C to prevent the huge losses and damages that we are now facing.'

I left Saleem as he headed into another meeting with his team. They would be at the COP to the bitter end. Rumours circulated that COP27 was to overrun by at least an extra day. Back at my desk, I continued working until the evening. Professor Jason Box had messaged to say he was introducing a screening of a new documentary called *Into the Ice*, a profile of his and his colleagues' dangerous but illuminating work on the Greenland ice sheet. In this documentary, while Box takes us across the skin of the changing ice sheet, his fellow glaciologists take us into the body of ice. They show us how it has been for thousands of years and how it is responding to the accelerating heating at the surface, through to where it calves icebergs into the Atlantic Ocean.

I arrived late at the Nordic Pavilion where Jason was readying himself to give his intro. There were several noisy events going on in this exhibition hall but the Nordic Pavilion had a sedate ambience that evening. There was also a good turn-out for the film and, as I arrived, I saw Professor Hugh Hunt and Dr Shaun Fitzgerald from the Centre for Climate Repair in Cambridge. Shaun was painting a steady blue line down one wall. The blue, soft-edged vertical lines, painted by visitors, covered the entire three walls of the pavilion and were part of a participatory artwork by Danish artist Jeppe Hein, titled *Breathe with Me*. Rugs and pillows were being laid out across the floor and it was getting cosy.

Jason hunted around to find a bottle with some wine in it, eventually splitting the contents of one between us. He nodded towards another attendee a few metres away who was standing at a forty-five degree angle to the ground, saying in a humorous tone, 'Yah, I guess he drank all the wine!'

For what my palate missed, the rest of my exhausted body was grateful. Jason has an interest in Austrian wine and as we chatted, a woman called Karen Bosman, who was with the South African delegation, interrupted us, saying, 'I'm from Stellenbosch!' which appeared to be the intro and the punchline rolled into one, before we were interrupted by the Nordic team, informing us that the movie was about to start. A chance meeting with Karen the following day led to a further source of information that would feed back into my still-emerging narrative from this COP.

The Overshoot Commission (reining in nature)

The next morning, I was back in the media centre early and preparing to head to two consecutive talks. I missed the first because I couldn't locate the pavilion but made it to the second, which was to introduce the Overshoot Commission into the flow of COP narratives. This presentation took place at the French Pavilion, given by the Paris Peace Forum, within which the Overshoot Commission was created.

The President of the Paris Peace Forum, Pascal Lamy, was fronting the esteemed panel to explain in very clear terms what the Overshoot Commission is tasking itself with achieving. 'We believe that, as science says, we will very likely overshoot 1.5°C; we have to re-look at the whole range of options,' he said. 'Mitigation, of course, remains the number one avenue and there is no discussion about this. I don't think anybody disagrees with that. Then we have to re-look at the three other ones: adaptation, carbon removal, and geoengineering, including solar-radiation modification.'

Lamy's statement on mitigation as the 'number one avenue' raises more questions than it allays concerns. With record-breaking global greenhouse-gas emissions year on year, with one or two exceptions, even a novice analyst would conclude that mitigation is the one avenue that has not been explored. Is this something the Overshoot Commission is going to correct by creating a radical plan to be adopted by the largest polluters and industry? He continued:

> This commission has as a purpose to table new recommendations, given that what happens between 1.5°C and 2°C is quite dramatic, new proposals to better handle adaptation, move up carbon removal, and look at the science and governance part of geoengineering, so that when we publish our recommendations, which is probably summer next year, we can reopen a public debate about these options so that they can be discussed, as we believe we should leave no stone unturned. That is the purpose of the commission.

What is apparent is that forcing developed nations and the high-emitting developing nations of the world to decarbonise is being left to the wholly inadequate COP process. The Paris Agreement was never enough to mitigate our way to 1.5°C or even 2°C. Many people I spoke to about this in the post-Paris euphoria raised their hands wisely and pronounced that we would meet the Paris obligations because, as nations, we would simply ratchet up ambition. The famous but largely unseen ratchet mechanism would come to our aid and bend the emissions curve to meet Paris. As Saleem pointed out shortly before, we have everything we need to do this, but we lack the political will.

Polls in the UK show that most of the population wants to stop all new fossil-fuel licences and speed up the growth of renewables, yet new licences for coal mines and North Sea drilling are still being issued. This is despite providing inadequate energy security to the British public. The politics are not only deceitful; they are out of step with the public mood.

Prior to this presentation, the Paris Peace Forum had hosted another panel less than a week before, also to present the Overshoot Commission. Lamy stated very clearly that the IPCC stated we would not meet the Paris limits for warming and that overshoot was certain. He restated the objectives of the Overshoot Commission: 'We have to re-open the concept that we have of adaptation, carbon removal, and geoengineering, i.e., SRM – solar-radiation modification.' What he said next showed subtle significance, but little self-awareness:

> I am now at the part of my life that is the end. There is a beginning and there is an end, as we all know, and I am at a stage where I want to leave no stone unturned. It wasn't that urgent when I was twenty years younger. It is becoming extremely urgent, and this is why I have accepted to spend time with people we've gathered and who *a priori* share this view that we have to re-look at that.

That any eminent diplomat might want to conclude their life with a legacy project is understandable. There are many libraries and even gigantic hydroelectric dams that carry the names of VIPs for this purpose, the world over. However, to state that the issue was not urgent twenty years ago, in 2002, can be interpreted as disingenuous because the Rio Earth Summit, where we were clearly warned that action must be taken, was in 1992, ten years earlier. COP1 took place in 1995 and the parties that attended were under no illusion why they were gathered in conference. The Kyoto Protocol was adopted in 1997 and entered into force in February 2005.

In 2002, Lamy was the EU Trade Commissioner and must not have been briefed that escalating carbon emissions were of any consequence. Between 2005 and 2013, he was the director-general of the World Trade Organization (WTO), where skyrocketing emissions might have at least once crept onto the agenda.

Listening to these presentations, it became apparent that the same people who wielded great power, with the ability to shape the world order, are the same people who sidestepped mitigation when it had been clearly stated that doing so would have consequences. Maurice Strong and Severn Cullis-Suzuki back in Rio, James Hansen in Congress, the sirens continued throughout the decades, accompanied by thousands of pages of IPCC reports stating the same thing: that global carbon emissions must come down or we would risk 'a moment in the twenty-first century where the condition of our species may become terminal'.

Today, the same people are telling us that mitigation has failed, that we must follow their advice further and look at ways to wrest control of the climate away from nature. The geoengineering question needs more open debate, but so does exploring and implementing far more radical programmes to significantly cut emissions. The IPCC does state that there is 'no credible pathway to 1.5°C in place' but it does not stop there. The UN report adds that it would require a 'rapid transformation of societies'. It is unlikely that the Overshoot Commission will waste any time on looking at what this rapid transformation of our societies would look like, even if the public were willing to get on board with the Marshall-style plan required to achieve it.

What the Overshoot Committee states as its objective is a set of recommendations that address adaptation, carbon-dioxide removal, and solar-radiation management. Jesse Reynolds, the executive secretary, is a geoengineering governance expert and, in the audience at the COP presentation, there were familiar faces from the climate-engineering research community, or *geoclique*.

As the talk finished, I came away thinking that the big names of former world leaders and diplomats were just part of the cosmetic dressing of an organisation with one purpose. Climate interventions have been very unpopular among many climate and social scientists, who have acted as a block to further research. The Overshoot Commission aims to provide the diplomatic muscle to run straight through the blockade.

The public debate should include all our options, but should primarily focus on making tangible what a 'rapid transformation of our societies' looks like, giving us the option to approve or disapprove of it. It is clear the COP cannot achieve this. It reminded me of what Kevin Anderson had been saying in 2017 at COP23 in Bonn: 'What we haven't tried in twenty-seven years is mitigation – actually reducing emissions!'

Five years on from Kevin's statement, the event at the French Pavilion demonstrated how seemingly disparate narratives appear at the COP and start slowly fitting together, resulting in a composite view of what the real picture is. The overdue rise of Loss and Damage, the ceaseless increase in fossil-fuel production, and the rising techno-optimism, pushed by a determined so-called geoclique, in order to solve the ills of civilisation with new frameworks for geoengineering research and eventual deployment, are what the COP is becoming. The theatre of the Sharm COP was much more polished because there was very little trouble from civil society barking at the gates, as in Glasgow. This was a fabulous stage to promote agendas.

I was standing outside the pavilion submerged in thought when Professor Hugh Hunt appeared and said goodbye. His time was up and he was heading back to the UK. As I waved him off, I heard a voice over my other shoulder yell, 'Hey, wine guy!' It was Karen from Stellenbosch. I asked what South Africa was up to at the COP and Karen explained about their plans to decarbonise the nation's current 85% dependency on coal for energy generation, in line with the Paris Agreement. They had developed a strategy to deal with the myriad issues, including a rapid reduction of the largest sources of emissions, retraining and relocating the people whose careers would be impacted by this, and also financing the transition, which involved support from larger countries.

I had been reading about the worsening droughts in South Africa and how extreme weather is having a calamitous impact. It is always impressive when smaller nations, who will struggle the most to adapt, let alone transition to clean energy, demonstrate leadership and take ambitious action. I asked Karen whether it would be possible to speak with someone in their team. After all, this was a positive story. My first at COP27.

South Africa had it very tough during the COVID-19 pandemic, where poverty caused enormous hardship and suffering. I interviewed a winemaker, Bruce Jack, at the time about how they were responding to the ban on both domestic and export sales of alcohol. The combination was decimating business resources. Bruce already had experience operating an NGO in the region and pivoted his business into a full-time relief agency, working with his staff to make sure people in the poorest areas had access to food and other supplies.

Speaking to Blessing Manale from the Presidential Climate Commission, I asked him how severe the climate impacts are that South Africa is experiencing and what they expected going forward. He said they have a two-fold crisis of prolonged drought in parts of the country and flooding in other parts. Their reporting shows that these are going to get worse. He stressed that their key priority was to make cities more resilient and to prepare for increasing exposure to impacts. They warn that: 'The cyclones will be changing; we won't be shielded by Madagascar any longer, so we are anticipating that in three or four years that Port Saint Johns and some of the beaches on the

East Coast will start to be hit. When Mozambique gets hit, we feel the pinch in Mpumalanga.'

The decarbonisation plan they have been developing is now being implemented. It marks a timeline out to 2050, which can sound like a long time but if the main source of emissions reductions is front-loaded into the plan by 2030, then this will be a remarkable achievement. The challenges that accompany this kind of transition are being faced in several developing and vulnerable countries, and South Africa's strategy is also feeding into the planning in some of these nations.

The willingness to discuss mitigation and build a credible pathway to reduce emissions already demonstrates more ambition than in many developed countries. The latter talk a lot about being leaders while sliding backwards on commitments, even when it is their own citizens being impacted by climate extremes.

Given the impacts South Africa was already experiencing and the scale of poverty and vulnerability in the country, I asked Blessing what he thought of the efforts in the COP27 talks to move the 1.5°C limit of warming. He told me:

> Once you say that we might overshoot the target, people are going to say, 'Well, we have lost it anyway. We don't want to go back to the days where the target keeps on moving. We are very far from Kyoto.'
>
> So we will continue to say, let's keep the target alive and work towards it, and where we miss it, we cannot blank it. We have to explain to our conscience that it happened because of external forces we could not account for, but we cannot do anything else.

Most of our consciences have been blind to the heating of the planet above the safe level scientists have been warning us about for decades. Of the 35,000 attendees here at COP27, how many are here to further an agenda that has little to do with reducing carbon emissions at the scale required? We are all failing at this. No one knows where we will be by the time our consciences finally grab us by the collar.

I worked until 5.00 p.m. and then headed back to the hotel feeling like an exhausted wreck. It was Friday evening, and the COP was to continue through Saturday for an as yet unknown period of time. The next morning, I went down to the pool and for the first time lowered myself into the cool water. It was 8.00 a.m. and the sun was already up over the cratered cliff that rose on the other side of the bulb-shaped pool. I swam the circumference a few times and then rested in the sun until the aroma from the chef cooking omelettes on the terrace convinced my stomach it was time to eat. In the absence of fruit, I had taken to eating a large plate of chopped raw cabbage, of which there was always an abundance on the breakfast buffet. Feeling the most refreshed I had all week, I packed up my things, checked out, and headed back to the COP. I planned to stay there until early evening and then take the shuttle to the airport for my return flight to the UK.

Arriving at the Blue Zone at COP during extra time – when the show has ended and the set is being deconstructed but certain essential facilities remain – was a surreal sight. The booths dispensing water and sponsored soft drinks were now unburdening their shelves to whoever wanted armfuls of Coca-Cola or Sprite. The media centre announced it would remain open for exactly twelve hours after the falling gavel officially brought the COP to a close. I returned to my desk for the last time and set to work editing and transcribing my material. I had arranged one last interview with Dan Bodansky at 3.00 p.m., for which I allocated an hour. Then I could pack up and depart.

The day sped through to 3.00 p.m., when I messaged Dan. He said he was in the Plenary Hall Ramses I, the furthest point away from the media centre within the COP. My Achilles tendon was tender and swollen, which had caused my limp to become more exaggerated. If I forgot about it for an instant and quickened my pace, I would be struck by a searing pain, bringing me to an anguished halt.

Entering the Plenary Hall Ramses I, it felt odd to be in a wood-panelled, marble-floored executive conference centre. I was much more accustomed to the bouncy floors of the temporary buildings. Through the banks of computers, between numerous small groups of people who appeared expectant, occupied in thought, or bursting to deliver a sliver of news to a colleague, I spotted Dan talking to a couple of people and waited to the side.

He finished and we walked out into a small terraced garden area with the last remaining food stalls still busy serving. A row of palm trees delineated the boundary. The sun was setting behind the palms and the mood was tranquil. Any evidence of the Global North/South divide was undetectable. It was time to probe Dan on his thoughts around the issues I had been contemplating and where he thought the road ahead might lead.

When we sat down on the last available table in this ethereal, landscaped corner of the COP, the sun was casting light-red to pink shadows across the scene, silhouetting the palms and sinking fast. In the course of our conversation, the hues turned to purple and blue.

I knew that in more recent years, Dan had been doing a lot of work in climate engineering governance. This was an opportunity to dig into the detail of climate overshoot through the legal lens. As a starting point, I wanted to hear Dan's perspective on why he was moving into this field of climate intervention:

> Even at 1.1°C, which is where we are now above pre-industrial levels, we are seeing massive harms from climate-change. It seems like we are on track to somewhere north of 1.5°C, maybe above two degrees. If there is a prospect of crossing tipping points, causing runaway climate-change, we need to know: are there other ways of trying to address that?
>
> I think it's prudent to be looking at climate interventions, climate engineering, because we may need them. They may not work, or they may be unsafe, so we're better off having done the research in advance. The worst thing would be to try something in desperation that actually isn't effective or causes catastrophic harm.

One scenario that is commonly highlighted is when a nation of any size decides that they want to take unilateral action to reduce the Earth's temperature. One method is to distribute a vast quantity of particulates into the lower stratosphere that would then be distributed around the planet and cause a reflective layer. This would mimic the effect of a large volcanic eruption, which has, in the past, shown a global cooling effect. The ensuing impacts of such an intervention are largely unknown, but climate models confirm that the intended outcome of cooling would be achieved. The trouble is, this is a blunt tool and no one knows how these changes would interact with existing components of the climate system, like potentially changing key weather systems like the Indian monsoon, or significantly thinning the ozone layer, for example.

I asked Dan if there was any legal framework or international process that existed, or was being considered for development, that could deal with this kind of scenario. He told me:

Generally speaking, there are not legal frameworks for many things that countries just decide as a matter of national policy that they want to do. I would hope that on something like climate interventions, where you are affecting the entire globe, you consult with others first. You try to make it as multilateral as possible. But in the end, there's unlikely to be a process that is able to make decisions about whether to use geoengineering. So if push comes to shove, I think countries will act, hopefully in collaboration with others and with consultation and transparency.

Dan agreed that it would likely be a unilateral decision by a nation, or a small group of nations, to proceed. I asked him what the wider response to this might be. He said:

The only existing institution that can address it is the Security Council, because the Security Council actually does have decision-making authority. I don't think there is going to be an ability to create another Security Council-like institution anytime soon. So, to the extent we may need climate interventions, then for the countries thinking about going ahead, I think the UN General Assembly and Security Council would be the forums where you would try to consult.

This is the horizon of where mitigation failure and climate engineering take us. We seem to discount radical societal change for a gamble on the development of technologies that may not work and, even if they do, have completely unforeseen outcomes. From conversations I have had with people working in the field, the only place to experiment meaningfully is in the atmosphere, at a scale that actually modifies weather. This sci-fi tech will need the same governance structures that currently fall into the hands of those who have mismanaged global affairs for so many years before. It takes a gigantic leap of faith to believe that this could have a positive outcome for those who see the COP as the only forum where they have a seat at the table.

If this was a serious forum with serious powers, the full focus would be on the elusive transformation of society. What does that look like? How do we transition at a much faster rate? What energy restrictions would we need to implement, and for whom? Would life be better or worse than it is now? Would future generations thank us? There are many of these questions, and the bulk of the answers lie outside of the high-security arena of the COP circus.

Climate engineering in the context that has been discussed is the contemporary equivalent of nuclear weapons, as nations calculate their own exposure to the crisis with diminishing interest in those who are facing it now. Researchers may well have good intentions, however, the inherent risks are not being discussed enough and by the people who are on the receiving end of the harshest and most rapidly approaching consequences of decades of ignorance. That we now must consider repurposing the UN Security Council, set up to confront aggression, to tackling climate disputes, is perhaps the clearest illustration of the failure of efforts to date.

Rabbi Yonatan Neril at COP27

Here are the words of Rabbi Neril:

'Every year, emissions go up. One definition of insanity is trying the same solution repeatedly and expecting a different result. The operating system that is driving humanity is consumerism. We need to change that operating system if we are going to continue to survive and thrive on this planet.'

Thirty years after Rio, when the world was unequivocally warned, the power in the world failed to do what was right to further a wider prosperity that extended beyond our species, to create a richer, more diverse and liveable planet.

Instead, the stewards of the world ignored the warnings in pursuit of a singular economic growth, ignoring metrics of well-being, over-consumption and the declining state of the global commons. The public has largely been misinformed, misled, misdirected, and today we are told it wasn't urgent then, but it is now, and it is too late to retain a safe climate.

Agency has been the key theme of all these COPs I have attended. It shifts ghost-like from year to year, group to group, and in the recent years it has been the noises outside the COP that have attracted the spirit of agency to create change. The public, in many of the high-emitting nations, now have the notion that we have to take action, but the agency to act still eludes them. The state, combined with powerful corporations and international diplomatic agencies, constantly reinforces the status quo with undertones of intimidation and condescension. In the UK, both main political party leaders are keen to lock up protestors for calling out their hypocrisy and lies, but when it comes to action to avert climate catastrophe, the problem gets a trim, and the situation worsens.

At the outset of COP27, the Prime Minister of Barbados, Mia Motley, expressed this sickness within our leadership very well, saying, 'But the simple political will that is necessary, not just to come here and make promises but to deliver on them and to make a definable difference in the lives of the people we have a responsibility to serve, seems still not capable of being produced.'

Motley's words should be remembered as we proceed from COP27 towards COP28. At the COP in Sharm El-Sheikh, there were 636 fossil-fuel lobbyists accredited, many affiliated with the world's largest oil and gas giants. That is more than any other single national delegation except the United Arab

Emirates, who are hosting the next COP. These lobbyists attend the COP to pollute the political process away from action that reduces the production and consumption of fossil fuels, towards one that prolongs their use, with dire consequences for all of us.

The fossil-fuel lobby was about to emerge from the shadows. In fact, they were to demonstrate just how much they had captured the UNFCCC COP process. The host nation for COP28 was announced as the United Arab Emirates and the COP President-designate was named as Sultan Al Jaber, CEO of the Abu Dhabi National Oil Company, ADNOC.

11

COP28, United Arab Emirates –

A Fossil-fuel *Fait Accompli*, or …

2023 was the year that felt materially different. Ocean heatwaves in the North Pacific and Atlantic coincided with the arrival of El Niño, a phenomenon creating above-average sea-surface temperatures that develop across the east-central equatorial Pacific. When I spoke to Professor Jennifer Francis at the Woods Hole Research Center in the US, she said, 'Expect chaos. Expect unusual events, expect extreme events. We're going to see heatwaves, we're going to see floods, we're going to see rapidly intensifying hurricanes. Those are all symptoms of climate-change.'

What ensued was a relentless barrage of extreme-weather events thrust upon a confused natural world. Near where I live, a pod of dolphins washed up on the shore at low tide. Over the hill in the next river valley, another pod was being carried back out into the ocean by local people. On a local surf beach, an exhausted basking shark became stranded and eventually died. In Brazil, on Lake Tefé, over one hundred Amazonian river dolphins and thousands of fish died as the heat of the water rose to what locals described as hot bathtub temperatures.

The animal die-off that struck me the most during the summer of 2023 was the estimated 10,000 baby emperor penguins that perished in Antarctica when warmer ocean currents caused the melting of the sea ice before the babies were sufficiently mature to survive. Emperor penguins are expected to have become extinct by the end of this century. Although I have never visited this region and have no reason to impose myself upon it, it has been through the films of Attenborough and others that there is a sense of familiarity and connection. Now we are seeing how continued burning of fossil fuels is awakening this colossus of the Earth's climate system.

Despite much of our mainstream media filtering out the suffering that people in poorer parts of the world are enduring due to climate heating, it was hard to ignore the estimated 19,000 British and other tourists fleeing the Greek island of Rhodes as dangerous fires swept through at high speed. In Canada, close to the Arctic Circle, forests burned so fiercely they coloured the sky in New York and sent a veil of smoke across the Atlantic to Europe.

The sheer might of nature was demonstrated in mainland Greece when 754mm of rain fell in just eighteen hours, creating lakes, destroying property and taking lives. A few weeks later, a second storm struck, exacerbating the suffering. In Libya, after the so-called medicane, a Mediterranean hurricane, struck the country, the death toll in the coastal city of Derna alone was over 11,000, with many people still unaccounted for. On 5 October 2023, Hurricane Otis appeared off the coast of Acapulco, surprising scientists and making rapid landfall as a Category 5 hurricane at speeds of 165mph (257km/h). Dozens of people died, and city-wide infrastructure was decimated. The local economy, largely dependant on tourism, will take years to recover.

Global Stocktake: a truly damning report card
Whatever your experience of the relentless impacts of climate heating to date, many different scientists are all telling me the same thing: it is going to get worse.

Amid all this, the Global Stocktake – 'a process for taking stock of the implementation of the Paris Agreement with the aim to assess the world's collective progress towards achieving the purpose of the agreement and its long-term goals' (Article 14) – was published almost unnoticed. It was summed up in a statement from Ani Dasgupta, President and CEO, World Resources Institute: 'The United Nations' polite prose glosses over what is a truly damning report card for global climate efforts. Carbon emissions? Still climbing. Rich countries' finance commitments? Delinquent. Adaptation support? Lagging woefully behind.'

Reflecting on how much of the suffering in far-off places is purposefully filtered out of the mainstream media and downplayed by policymakers, we can start to see how *uncaring* is instrumental to reinforcing the status quo. By concealing reality and misinforming the public, our concern is dampened and those calling for justice, action and adaptation have their voices silenced. In our ignorance, we remain unprepared.

The much-needed funding for the transformation of society, even in places with great wealth, as in the West, is siphoned back out as subsidies to fossil-fuel companies and similar zombie enterprises, sustaining a system that is economically and morally bankrupt.

Professor Julia Steinberger: 'He's an oil man, that is literally what he is.'

The Global Stocktake was set to be a central theme of COP28 UAE, overseen by Sultan Al Jaber. Al Jaber, as COP President and Abu Dhabi National Oil Company (ADNOC) CEO, was busying himself trying to balance the conflicting position of saying he is going to reduce carbon emissions while having a clearly stated goal to expand ADNOC's oil and gas production by an estimated 7.5 billion barrels over the next few years.

When I asked Sir David King, chair of the Climate Crisis Advisory Group, what he thought about Al Jaber's position, he said, 'Simply having COP28 in the Middle East, the biggest oil-producing area in the world, was already a red flag. Sultan Al Jaber believes we will consume more and more oil and gas globally and we will capture all of the carbon dioxide to get to net zero by 2050. So there's a person who's in the presidency who is the nightmare-scenario person.'

This sentiment was echoed by Dr Joe Romm, former Assistant Secretary of the US Department of Energy during the late 1990s, when he said, 'We are now at a serious point. Everyone can see there are massive heatwaves all over Europe. There has been a record twenty days in a row in Phoenix of above 120°F [49°C] temperatures. The temperature in Death Valley at midnight was 100°F [38°C]. We are warming up. We are past the point of these sorts of games and we should not have let an oil company do this sort of thing.'

The problem is, the United Arab Emirates (UAE) is not the only country looking to expand its national fossil-fuel-production capacity. When I interviewed the chair of the Fossil Fuel Non-Proliferation Treaty Initiative, Tzeporah Berman, she said Prime Minister Trudeau's climate record was scoring well on all the things that we measure climate policy on. Yet emissions are still going up.

Oil and gas in Canada is the fastest growing sector, not from domestic heating or transport but from the production and expansion of fossil-fuel products. Tzeporah says this is because oil and gas production is not a criterion by which climate leadership is judged, despite it being a root cause of the climate problem.

She said, 'This locks in future emissions. You don't build a pipeline for ten years. You spend twenty billion on a pipeline, like Trudeau did, thinking that forty or fifty years from now we are going to be consuming more oil. The theory is that demand will go down and then, magically, supply will go down, but it is not happening.'

I had a call with Professor Terry Hughes at John Cook University, who is also the Director of the Australian Research Council (ARC). We discussed the increased bleaching of the Great Barrier Reef that is killing this massive but sensitive ecosystem. Coral reefs are what we call 'keystone species' that support whole webs of marine biodiversity. Year after year of ocean heating leaves no time for recovery. Terry links this directly to the burning of fossil fuels, which in turn warms the atmosphere and oceans. Commenting on the continued expansion of coal production in Australia, he said:

> Here in Australia, people voted to change the government with the expectation that there would be a marked change in energy and emissions policy. Unfortunately, many of those voters have been disappointed. Support for fossil fuels remains strong within both major political parties. I think politicians are aware of growing public support for renewables. When a permit for a new coal mine is refused, it is usually accompanied by a big press release, but when many more are approved, it's done at a quarter to five on a Friday when no one is there to report on that story.

In the UK, the government approved the permit for the Rosebank oil-drilling project in the North Sea, which it is estimated will extract 500,000,000 barrels of oil. The opposition Labour party has signalled they will not revoke the licence if elected. Deceitful claims about economic necessity and fuel security are easily debunked. Responding to the government's false claims, environmental lawyer Tessa Khan said on Sky TV:

That is simply not credible. We heard yesterday from the International Energy Agency that we can't have new oil and gas fields being developed if we are going to stay within the critical threshold for a safe climate, which is 1.5°C. It is also not what is needed for energy security or affordability. We know that eighty per cent of the oil being produced in the UK is being exported and sold on international markets. That oil and gas will not bring down household energy bills, but it does entail us giving a £3.75 billion subsidy to the oil giants, who will be making a £1 billion profit to develop this field. It is a disaster for the climate, for our economy and for our energy security.

Add all this together with the many other nations, including Saudi Arabia, Norway and the USA, and there is no such thing as a carbon budget any more than there is any such thing as hope for a stable future as the world transforms into a hellscape around us.

All these countries promise that they are cleaning up their fossil-fuel production and aligning with net zero. Al Jaber says ADNOC will invest in carbon-capture technologies that don't currently exist on a scale anything like what would be necessary. There is the evolving scandal around carbon credits, where countries and large corporations pay, usually poorer nations, by the tonne to store carbon in a multitude of ways. The purchaser then uses this to offset against their own national carbon accounting. It is widely dubbed the 'licence to pollute' and shifts the actual work of reducing emissions to a future date at a higher cost.

Former US Vice President Al Gore has stated that Al Jaber's position at the COP is unacceptable. Both Gore and former UNFCCC Executive Secretary Christiana Figueres, have said that the fossil-fuel lobbyists should be removed from the COP. Figueres wrote in an op-ed for Al Jazeera: 'The industry as a whole is making plans to explore new sources of polluting fossil fuels and, in the United States, intimidating stakeholders who have been moving towards environmental, social and governance responsibility.'

COP28: a scandalous desert shindig

On the eve of COP28 the first of several scandals broke. The first was reported by the Centre for Climate Reporting which had obtained leaked documents showing that COP President Dr Al Jaber was planning to use the conference to hold closed-room meetings promoting oil and gas deals. He denied the allegations. This was quickly followed by another scandal, also uncovered by the Centre for Climate Reporting and Channel 4 News in the UK. They published evidence of a plan spearheaded by Saudi Arabia's Crown Prince Mohammed bin Salman, to artificially stimulate oil and gas demand in Africa by implementing around fifty projects that would lock-in ongoing fossil-fuel use for decades to come in defiance of global efforts to reduce dependency.

As the world looked on in dismay, a third scandal broke, centred once again on COP28 President Al Jaber, who had said to former President of Ireland, Mary Robinson, that there was no science that stated fossil fuels needed to be phased out to limit warming to 1.5°C. He went on to say that phasing out fossil fuels would limit sustainable development 'unless you want to take the world back into caves'. UN Secretary-General, António Guterres, said Al Jaber's words were 'very troubling' and 'verging on climate denial'.

Public perception of the desert shindig was getting more cynical by the moment. A week later Al Jaber was forced to call a press-conference where he catalogued a list of 'historic' achievements that had already occurred. However, when questioned, he appeared churlish, suggesting he was being victimised by the press and that his words were being taken out of context. To his right sat the chair of the IPCC, who then endorsed Al Jaber's understanding of the science, saying that in 1.5°C-compatible scenarios 'by 2050, fossil-fuel use is greatly reduced and unabated coal use is completely phased out'. He then stated oil use by 2050 is reduced by 60% and gas by 45%.

When I spoke to energy and climate expert, Professor Kevin Anderson, shortly after the press-conference, he said, 'The ridiculous statement that comes out of the chair of the IPCC in defence of the chair of the COP, that we can have 1.5°C with no more coal but lots more fossil fuel in 2050, simply does not fit with the maths. This is because the chair of the IPCC has been, I was about to say hoodwinked, but he is fully aware of this so it is not hoodwinked. He has been completely taken in by the narrative around negative emissions technology.'

These geoengineering forms of direct air capture and other methods to try and remove carbon from the atmosphere remain in their infancy, but it is being assumed by fossil-fuel-producing nations that they already work at scale. As has been repeated throughout these pages, despite the billions invested, they do not work at anything like the scale needed and it is not at all certain that they ever will. It is clear that they are being used as a fig leaf to cover the dangerous intent of nations like Canada, USA, UK, Norway, Saudi Arabia, UAE, Australia, Russia, China and so on, to destroy the only atmosphere we have for a few dollars more.

Exotic cars and neon nightmares
It was into this fray that I had arrived at the tail-end of week one in Dubai. The 3.00 a.m. journey from the airport to my hotel was painless. My electric taxi buzzed along the long stretches of asphalt multi-lane highways, that are the concrete arteries of this neon nightmare of a city. The roads were largely empty,

except for showrooms stuffed to the plate glass with 'exotic cars' and other trinkets. Although I fell asleep at 4.00 a.m. I was up at 9.00 a.m., clambering to my feet to get to an appointment in the Green Zone.

Outside the air was thick and the pervasive haze rubbed out the hard edges of the towering glass structures appearing over the other side of the raised metro tracks. A guy stood next to a Lamborghini, its door up and its engine idling, smoking, a twenty-first century Marlboro man.

I walked towards the station and immediately saw three garishly coloured Ferraris banked on the curb. A giant hearse of a Rolls-Royce, ugly as sin, parked around the corner. By contrast, the metro is a haven, set apart from the space-junk consumerism of the streets. The sedate bubble of relative normalcy is burst when, out of the haze, the vast structures of fossil-fuel refineries which dominate the landscape appear, and a chimney flaring 'natural' methane gas completes the scene.

My meeting with Jack Curtis from Carbon Jacked is constantly interrupted by giant helicopters landing and taking off, ferrying VIPs into the enclosure. Afterwards, walking around the enormous complex, it struck me that this was no ordinary COP. Its great size, giant permanent buildings, and glossy interiors may as well be a spaceport to another universe. The Saudi Green Initiative being the most audacious of them all. A dizzying immersive experience of false futures that are wholly dependent on continued pumping of oil and fantasy air-scrubbing technologies. I speak to a young engineer about his model of a carbon-capture and storage (CCS) plant. At scale it would have the capacity to suck five hundred thousand tonnes of carbon dioxide from the atmosphere per year. I ask how they are going with scaling them up. He smiles. 'This is just a pilot. It will only be scaled up if *they* want it.'

'Who is *they?* I ask. He looks awkward. 'Your government?' I say, to which he replies with another awkward gesture to the affirmative. 'What would *make* them want it?' I say.

'If the expansion of petroleum sales continues.' He says.

So no efforts will be made to capture current emissions, which in the past year are estimated at 608 million metric tonnes from 'industrial processes'. None of this includes emissions from where the fuels are ultimately used. No fossil-fuel producer includes that in their budget. I smiled gratefully at the polite young man and walked on.

This was the COP with the largest ever attendance: around 100,000 registered delegates. Despite this, the pattern was the same as ever. News broke of tensions in the negotiations. Everything began to hinge on the phrases 'phase down' and 'phase out'. The former was weakened language in a formally negotiated agreement which would allow countries to go on burning fossil fuels indefinitely. The second was a definite term that literally meant *stop.*

The media centre was abuzz with chatter about villainous countries that were plotting to make sure the text was weakened. Others were adamant that we would get the ambitious wording that the world needed, as if those different words would make even the smallest change to policy concerning fossil-fuel production in the most extractive nations.

The farce continued. UK Prime Minister, Rishi Sunak had flown in as the ink was drying on new fossil–fuel extraction licences on his desk back in London. John Kerry was spotted on numerous occasions swanning around in the back of golf carts, as if on holiday. As the talks seemed to be going nowhere, the tension rose. There was the air of a predictable ruse. Let the talks look like they are going to fail. There will be chanting, outcry, protest-lite (those of the estimated 2,456 registered oil industry executives and lobbyists probably had a good chortle as they passed by), and then Al Jaber would show his mettle, talking tough as if about to bang some heads together. Kerry would stay up all night to rough-feel some collars, and in the morning we would get wording in the agreement that was better than what might have been and therefore a triumph for all. See you next year.

All the cards on the table...

At COP27 the Overshoot Commission had presented themselves as a high-level independent organisation looking at controversial geoengineering techniques to cool the surface of the Earth. Their report released in the summer of 2023 garnered some press but any attention fizzled out. The report called for an international moratorium on geoengineering, but allowed for research to take place.

I was curious to see if there would be any resurgence at COP28 but there was very little activity. People I spoke to agreed that the commission came across as ill-conceived and top-down in its over-use of former occupants of high office.

That was not to say the climate intervention, or *geoengineering* space was dead. Far from it. I was invited to a session being hosted by the Washington-based Silverlining Institute, titled 'No 1.5°C Without Climate Intervention' in the Climate Live pavilion. The red exhibition space branded with 'Climate Live' repeated all over the walls looked more like a trendy crypto event than a climate presentation. The space was tightly packed with people crammed onto six rows of six seats. A young Finnish woman called Anni Pokela, representing a group called Operaatio Arktis, was on stage going through a climate-science elevator pitch spelling out that we are overshooting the 1.5°C boundary for global heating that scientists warn we should not cross.

Looking to my right I noted Professor Dan Bodansky, the esteemed climate lawyer from Arizona State University, tapping notes into his laptop. The science wasn't new but the vibe was different. Usually I saw young people at these events angrily denouncing research into climate interventions. Anni was talking them up.

As she took her seat, she was replaced by a panel of four young speakers from the Global South. One, a young guy called Ricardo Pineda from Honduras explained how hurricanes, made much stronger by the added moisture from a hotter world, had caused so much damage in his home country that development had been set back by twenty years. These impacts, made worse by the carbon emissions of the wealthiest people in the wealthiest nations, were setting places like Honduras on an accelerating pathway to annihilation. Ricardo went on to say that there was no Honduran representation in scientific organisations like the IPCC. Many other nations like his remain outside the scientific dialogue on climate action. He was adamant that research into geoengineering that might help save lives and infrastructure at home, must be financed and scaled up. He said that to take those cards off the table, as many Global North climate scientists were saying we should, would be nothing short of a human-rights violation.

Joshua Amponsen, from Ghana, founder of the Green Africa Youth Organisation, added to what Ricardo was saying, stressing that in his country they had one woman physicist contributing to the IPCC. He said that this one woman was why many other young people in Ghana were getting inspired by science. The trouble is that they quickly realise they have limited resources and relocate to wealthier countries where there are greater opportunities. This brain drain of early-career academics reinforces the vulnerability of these nations. Joshua was arguing for greater representation, access to supercomputers and for more responsibility to be given to African researchers to lead in research publications. When it came to climate interventions of the geoengineering variety, he doubled down on what Anni and Ricardo had said.

What interested me about this was that for the last decade or so, there had been many arguing that geoengineering was an extension of colonialism: a Dr Strangelove technology that would be imposed on the Global South nations, increasing risks from unknown consequences, adding to the climate impacts already being endured. These new voices were inverting that message by saying, 'You've already missed all your own targets, broken your own promises, and now you want to exclude us from deciding what options are available to explore!'

One polar researcher who is against these interventions is Dr Heidi Sevestre, affiliated with the Arctic Council. When I asked Heidi what her response was to the kinds of climate interventions being talked about in numerous sessions around the COP, she expressed great concern that the scientific community was being used to justify positions that were not reflected in their research. Importantly, she said, 'We know that solar-radiation management (SRM) could, for example, create more acid rain, or could disturb the Indian monsoon. It would be the whole water cycle that would be affected.'

Heidi went on to counter the argument that geoengineering could buy us more time to respond to the impacts of climate-change by saying that progress is too slow to be able to develop them in time. I did wonder whether if we had began researching them properly a decade ago might we not now be in a better position? However, the same could be said for phasing out fossil fuels.

What struck me most forcefully was that this conversation was entering a new phase. The many years of repetitive, deadlocked debate that had effectively blocked research was being overturned by the awareness of extreme climate impacts coupled with a new generation of global voices who are inserting themselves into the discussion with highly articulate arguments. Global equity and the inclusion of diverse voices is often heard as a central demand in making progress on climate action. But what happens when we in the Global North don't like what those Global South or youth voices are calling for?

The belly of the beast

Walking back to the media centre I noticed a growing number of people had started to gather by the entrance. The first draft of the COP28 agreement had been released and a few analysts were about to start a press briefing. Just as I approached, another fifty or so journalists appeared, creating a swarm-like effect. The message was quickly deciphered. In line with the art of twisting plots, the COP presidency had allowed a text to be released that had actually removed both 'phase out' *and* 'phase down'. As one analyst said, 'Phase out has been phased out!'

The mood sunk low. This was the nightmare-come-true of having the COP in an oil state. There was outcry and anger across the board. Al Gore raged, Christiana Figueres was incensed. The chanting from the designated protest space grew louder and there were calls for Al Jaber's resignation. By morning the office of the COP presidency had published a statement: 'As you know, yesterday we released a text. As you also know, lots of Parties felt it did not fully address their concerns. We expected that. In fact, we wanted the text to spark conversations … and that is what happened.'

When I spoke to Tzeporah Berman from the Fossil Fuel Non-Proliferation Treaty Initiative, she was upbeat because for the first time in thirty years, a COP text had actually mentioned the words *fossil fuels*. Amazingly, in the three decades stretching back to the Rio Earth Summit of 1992, the root cause of the problem had never been mentioned. It was blocked by polluting nations from even appearing in the text. She said, 'Before this text came out, the majority of countries on the floor here said they would support a phase-out of fossil fuels. This draft was written by OPEC.'

A couple of days previously it had been leaked that the head of OPEC had written to heads of state in an attempt to warn them off focusing on fossil-fuel phase-out. This was an unprecedented move by a powerful member of the fossil-fuel lobby. Tzeporah exuded confidence, saying, 'It shows they're terrified. They're terrified of the conversation around fossil-fuel production. And in the memo, they said you cannot talk about production, you must just talk about emissions.' She called this the beginning of the end of fossil fuels.

I asked Tzeporah if she thought hosting a COP in the UAE was a good thing and she replied, 'It's great. Bring it on. We've been pretending that we can have a conversation by promising these technologies that are not viable at scale and planting more forests, and therefore we can just keep approving more oil drilling, fracking and coal mines. Well, it's not working. Every year, emissions go up instead of down and every year, we wonder why it's not working. Economists would say we're trying to cut with one half of the scissors: just the demand but not the supply.'

All this was in the cold print of the United Nations Environment Prográmme Gap Report that was published stating that globally we are on track to produce 110% more coal, oil and gas than we can ever burn if we are to keep temperatures at a safe level. With all the major fossil-fuel economies expanding production, we are going to overshoot this figure by a long way and make the planetary situation far worse. Until we can halt supply, the COP outcome will continue to be a charade acted out by one group of very powerful high-emitters, deceiving us as the state of our world worsens before our eyes.

As Kevin Anderson had said in our interview a few days before, 'The senior people at COP, the government officials and CEOs of oil companies, these are not representatives of our country. They are much more closely linked to each other than they are to their nations. Their nations are a secondary place. These are, effectively, international citizens. There is a particular group of them, and these are the ones who have done remarkably well. I don't just mean well-paid professors, I mean a group of people that are just stratospherically in a different place from their income, power, wealth and resources. Sunak, Kerry, the chair of the COP, these are not people from different countries. They are all from the same country and that country is called Obscene Wealth.'

As the negotiations went into overtime and tensions were running high, I packed my bags and prepared for the long journey back home through the night. Walking to the exit I saw a familiar face and waved. He waved back, saying, 'Maybe see you in Baku!' It had been announced that COP29 was to be held in Azerbaijan the following year. The implications of another nation heavily economically dependent on fossil fuels using their position as host to massage an outcome they would be happy to sign up to, feels like a death knell to any idea of real progress in tackling climate chaos. The fossil-fuel industry are making a mockery of global efforts to build a cleaner and liveable world.

The eventual text that came out of COP28 was hailed as an historic agreement by Dr Al Jaber himself, but characterised as weak in terms of getting us anywhere near achieving bending carbon emissions towards the 1.5°C warming boundary. In fact, we are apparently crossing that boundary as I write, entering the overshoot stages of global heating. To pull it back, we need an aggressive restructuring of society. Many of us will witness enormous losses first hand. The future of climate negotiations is going to get tougher because we will no longer be arguing with oil and gas executives; we will be contesting with Nature herself. The consequence of so much greed and consumption will be an era of breached tipping points. We will need to hold on for dear life.

To underscore the lunacy of the COP, a few days after agreeing to transition away from fossil fuels, Al Jaber resumed his role as CEO of the Abu Dhabi National Oil Company (ADNOC), telling the *Guardian* newspaper, 'The world continues to need low-carbon oil and gas and low-cost oil and gas.' ADNOC are planning to invest $150 billion in more oil and gas extraction over the next seven years.

It is crystal clear beyond all doubt that we do need a Fossil Fuel Non-Proliferation Treaty to be adhered to. We need nations, such as the UK, which are wealthy enough to transform their societies to renewables, to send the signal that fossil fuels are undergoing a rapid phase-out. Such a signal would reverberate around the world, creating momentum for other nations to follow. We must question the motives and the psychology of the people pursuing fossil fuels in the face of so much scientific evidence that we are heading straight towards catastrophe.

Many of us cite the young as an inspiration for caring about the planet, quite often as a proxy for our own lack of care. When I spoke to Sally Weintrobe, author and long-standing member of the Climate Psychology Alliance, she emphasised that we struggle between a part that cares and a part that does not care. She highlighted that our care is powered by feeling entitled to life and to protect life. This 'lively entitlement' describes the psychological motivation towards a wider perspective that includes the self as part of the collective. By contrast, uncare is powered by feeling 'narcissistic entitlement' to claim all for self and in-group and to disregard any reality that limits greedy acquisition. Although many of us can admit to a degree of narcissistic entitlement, like pursuing goals that flatter our egos, it is the space between these two forms of entitlement that determines the level of guilt we are experiencing when we see suffering increase and the natural world in sharp decline.

However we approach the climate issue in our individual ways, it is time now to constantly question the motives of our leaders, our peers, ourselves and others. Are we acting in a caring way, with characteristics of lively entitlement? Or are we stuck in patterns of narcissistic entitlement, produced in part by the modern age, that we need to move away from? None of us can be perfect, but we can be better in our commitment to acknowledge and take appropriate actions on a scale that befits us. We can become politically engaged; we can adapt our lives as much as possible to reduce carbon emissions. We can also show empathy to people around the world who are suffering and extend that empathy to other creatures.

At a time like now, when politics is ranged against life on our fragile planet, it is easy to succumb to depression. Professor Julia Steinberger paraphrased Gandhi, saying, 'First they ignore you, then they laugh at you, then they fight you, and then you win. The ignoring thing happened. The laughing at you with overt climate denial happened. *This is the fight, right?* This is an overt fight.'

Courage, *corage, cœur*, the heart
After nearly three decades of cop-outs, it is no longer possible to deny that our addiction to burning fossil fuels has caused irreparable harm to the global commons. Forests are burning, food systems collapsing and our future on this path can be only bleak.

Over a decade ago, concern about climate-change was a taboo, eccentric issue to be aligned with. However, people all over the planet are being confronted by climate risk every day and waking up to the perilous situation we are all in. If you have read this far, perhaps you are concerned too. In this broad coalition of people from all the regions of the world, with different cultures and levels of resources and agency, we have to ask: what are we going to do about it?

Searching for profound and serious words of advice makes me reflect on the last fifteen years of interviewing people about this. I would like to give the last word in the book to someone whom I greatly admired. The late environmental barrister, Polly Higgins, made it her life's work to campaign to create a law of ecocide that would make harming nature a crime against peace. The campaign is as vibrant and as necessary today as it was when Polly was alive and campaigning.

When we last met in 2014, she was at her home in North London, packing up to move to Stroud, Gloucestershire, in England. After my interview, Polly asked if I could turn the camera on for her to make an additional appeal to people and for it to be played wherever I thought it might be relevant. In this moment of great uncertainty and, in terms of human spirit, here are Polly's words:

This is my invitation to you to dare to be great, and I say that to absolutely everyone. This is a moment in time, in the whole history of civilisation, when we are being called upon to stand up and speak out as voices for the Earth. And by greatness, I mean being in service to something greater than the self.

We can choose a different pathway. We can choose to go a different way in our lives, but it takes courage to do that – *corage, cœur*, the heart. It is the age of the heart, if you like. So it is coming from the heart. It is not coming from the head. Greatness that comes from the heart is something far greater than the self. It speaks of standing up and speaking out, even if it feels that it is getting us out of our comfort zone, even if it feels as if the whole tide is against us, because each time we do that, and we do it from a place of deep care, we give permission to others to do the same as well.

This is what I'm doing with a law of ecocide. I am standing up and speaking out from a place of deep care to create a legal duty of care. Because also, care and greatness are deeply, intimately connected. It is care that will create the greatness within each and every one of us to take the next step into the unknown and make this world a better place.

Thank you.

Acknowledgements

A debt of gratitude is owed to a great many people – far too many to mention – who have contributed thoughts, insights, encouragement, support and often a great deal of time.

I want to mention those who predate any first inspirations to write COPOUT. It has to be said that the most formative person in this whole expedition into reporting on climate is the late Jim Lovelock. I would not have met Jim if it was not for my very talented friend, Andy Worboys, who agreed back in 2009 to come with me on a journey into Jim's Gaia world that ultimately expanded my perception of our delicate planet. Meeting and interviewing Jim again in 1999 on the eve of his hundredth birthday was another milestone in my allotted years of trekking across Gaia.

The idea for writing this project was inspired by two conversations just after the COP26 in Glasgow. The first was with the author Tom Rosenstiel. Tom was asking what my perception of the COP was, and after I replied giving the context of the previous COPs going back to Paris, he suggested that, 'Someone needs to write this up in a gonzo style.' About a week later, I was talking to my old friend Richard Payne, who serendipitously said, 'You should write a gonzo account of the COP.' I am not sure if one of the suggestions alone was enough, but two tipped the balance.

Over the last fifteen years, I have conducted hundreds of interviews with climate experts. I doubt there is a single conversation where something wasn't gained, be it new insight, something to reflect on, or a point to push against. In the spirit of trying to report with transparency and fairness, I owe a great debt of gratitude to every single interviewee. There are many that I find it hard to agree with but those often provide the most cause for reflection, so thank you all.

I would like to offer a special thanks to Kevin Anderson and Saleemul Huq, two scientists that I have interviewed multiple times and who have helped me articulate certain complex climate issues that, as Kevin would say, emerge out of the science. I also owe a great deal of gratitude to Hugh Hunt, who has been a good sounding board on all things climate and engineering over the last decade or more and has given his time generously to me and to many others on similar knowledge-seeking journeys. There are quite a few among the interviewees that I now regard as friends. Private conversations have been as informative as the interviews, as much for the reminder that we are all people trying to make sense of this world and ensure an enduring comfort for our communities and other species going forward.

I also want to express gratitude to Mike Coe and Lizzie Stoodley, who attended COP21 to help with the filming relating to myself and Scott and also assisted with other filming projects. In the post-Scott era, I would also like to extend thanks to Heidi and Charles Brault, who have continually been on hand behind the scenes, organising events at successive COPs.

Aside from the many interviews that inform this work, it has also been the emergence of activists all over the world that has provided the glue and the fuel for progress. Many of these activists put themselves in harm's way or are facing highly punitive actions from governments who are not willing to listen. Protest is a function of a healthy society, especially at times like this when it needs to address itself and course-correct. I owe a significant debt of gratitude to all past, present and future people taking a stand against injustice.

I would not have got this far if it were not for my friend Sunny Singh introducing me to my agent, Tom Cull. Tom has worked wonders, first guiding me towards a rigorous editing process that has been as exhausting as it has been rewarding. All the remaining blemishes and imperfections are one hundred per cent my own. Additional thanks to Tom for doing what every author hopes an agent will do: finding a publisher.

I am indebted to *The Ecologist* for providing all my COP passes over the years. Without them, I would not have been there. I am especially grateful to Crispin Tickell and Brendan Montague and also to Melanie Hewitt, who read the first draft and was very encouraging.

A big thank you to the team at Ad Lib Publishers and my editor, Duncan Proudfoot, who has, ever so politely, pushed me further along the road towards a finished item as best as I can manage.

Lastly, this could not have been achieved without the ongoing support and affection of my partner, Natalia Baloghova. The first draft was written during the incredible heatwave period in Italy in 2022. In order to sustain productivity, I had to become nocturnal. For a while, we were like ghosts passing in the heat. Accept my heartfelt gratitude for your unwavering love and all the time we spend together.

Index